Walter Dwight Wilcox

Camping in the Canadian Rockies

An Account of Camp Life in the Wilder Parts of the Canadian Rocky Mountains

Walter Dwight Wilcox

Camping in the Canadian Rockies
An Account of Camp Life in the Wilder Parts of the Canadian Rocky Mountains

ISBN/EAN: 9783337096540

Printed in Europe, USA, Canada, Australia, Japan

Cover: Foto ©Andreas Hilbeck / pixelio.de

More available books at **www.hansebooks.com**

CAMPING IN THE CANADIAN ROCKIES

AN ACCOUNT OF CAMP LIFE IN THE WILDER PARTS OF
THE CANADIAN ROCKY MOUNTAINS, TOGETHER
WITH A DESCRIPTION OF THE REGION ABOUT
BANFF, LAKE LOUISE, AND GLACIER,
AND A SKETCH OF THE EARLY
EXPLORATIONS

BY

WALTER DWIGHT WILCOX

WITH TWENTY-FIVE FULL-PAGE PHOTOGRAVURES, AND MANY TEXT
ILLUSTRATIONS FROM PHOTOGRAPHS BY THE AUTHOR

G. P. PUTNAM'S SONS
NEW YORK   LONDON
27 West Twenty-third Street   24 Bedford Street, Strand
The Knickerbocker Press
1896

The Knickerbocker Press, New York

PREFACE.

THE Canadian Rocky Mountains offer exceptional attractions to those who enjoy natural scenery, sport, and camp life. Few regions of the world combining mountain, lake, and forest scenery possess the additional advantage of a delightful summer climate, such as obtains in the Canadian Rockies.

The extremely wild character of this part of the Rocky Mountains, and the very short time since it was opened up to travellers, are probably, in great part, the reasons for the lack of literature and the absence of any thoroughly illustrated publication concerning this region.

During a period of four years, the author has made camping excursions into many of the wilder parts of the mountains and effected a considerable number of ascents. An excellent camera has been an almost inseparable companion in every excursion, so that photographs of the typical scenery have been obtained from every possible point of view. Moreover, throughout all the processes of photographing, no expense of time or labor has been spared in order to obtain true and artistic representations of nature. Nor have these results been obtained without considerable sacrifice, for in many cases the proper light effects on lakes and forests required hours of delay, and frequently, on lofty mountain summits, high winds made it necessary to anchor the camera with stones; while the cold and exposure of those high altitudes made the circumstances unfavorable for successful work.

A map is not included in the volume, as, owing to the wildness of the country, there are no detailed maps covering this region that are entirely satisfactory. The best map, and, in fact, the only one available, is published in Dr. Dawson's *Preliminary Report* on this part of the Rocky Mountains.

The author makes grateful acknowledgment of the assistance received from many friends in the preparation of this book. Special thanks are due to Prof. J. H. Gore, of Columbian University, and to the Hon. Chas. D. Walcott, Director of the United States Geological Survey, for the valuable aid and information given by them; to M. Guillaume La Mothe for an interesting letter concerning the first exploration of the Fraser River; and to Sir William Van Horne for the many courtesies extended.

W. D. W.

WASHINGTON, D.C., July, 1896.

CONTENTS.

CHAPTER I.

Banff—Its Location—The Village—Tourists—Hotels—Topography of the Region—Rundle and Cascade Mountains—The Devil's Lake—Sir George Simpson's Journey to this Region—Peechee the Indian Guide—An Indian Legend—The Missionary Rundle—Dr. Hector—The Climate of Banff—A Summer Snow-Storm—The Mountains in Winter   1–15

CHAPTER II.

Lake Louise—First Impressions—An Abode of Perpetual Winter—The Chalet—Visitors—Stirring Tales of Adventure—Primeval Forests—Forest Fires—Mosquitoes and Bull-Dog Flies—Mortal Combats between Wasps and Bull-dogs—The Old Chalet—Morning on the Lake—Approach of a Storm—Sublimity of a Mountain Thunder-Storm—Cloud Effects—The Lake in October—A Magnificent Avalanche from Mount Lefroy—A Warning of Approaching Winter .   .   .   16–35

CHAPTER III.

Surroundings of the Lake—Position of Mountains and Valleys—The Spruce and Balsam Firs—The Lyall's Larch—Alpine Flowers—The Trail among the Cliffs—The Beehive, a Monument of the Past—Lake Agnes, a Lake of Solitude—Summit of the Beehive—Lake Louise in the Distant Future   .   .   36–46

Contents.

CHAPTER IV.

Organizing a Party for the Mountains—Our Plans for the Summer—William Twin and Tom Chiniquy—Nature, Habits, and Dress of the Stoney Indians—An Excursion on the Glacier—The Surface Debris and its Origin—Snow Line—Ascent of the Couloir—A Terrible Accident—Getting Down—An Exhausting Return for Aid—Hasty Organization of a Rescue Party—Cold and Miserable Wait on the Glacier—Unpleasant Surmises—"I Think You Die"—A Fortunate Termination 47-64

CHAPTER V.

Castle Crags—Early Morning on the Mountain Side—View from the Summit—Ascent of the Aiguille—An Avalanche of Rocks—A Glorious Glissade—St. Piran—Its Alpine Flowers and Butterflies—Expedition to an Unexplored Valley—A Thirsty Walk through the Forest—Discovery of a Mountain Torrent—A Lake in the Forest—A Mountain Amphitheatre—The Saddle—Impressive View of Mount Temple—Summit of Great Mountain—An Ascent in Vain—A Sudden Storm in the High Mountains—Phenomenal Fall of Temperature—Grand Cloud Effects, 65-83

CHAPTER VI.

Paradise Valley—The Mitre Glacier—Air Castles—Climbing to the Col—Dark Ice Caverns—Mountain Sickness—Grandeur of the Rock-Precipices on Mount Lefroy—Summit of the Col at Last—A Glorious Vision of a New and Beautiful Valley—A Temple of Nature—Sudden Change of Weather—Temptation to Explore the New Valley—A Precipitate Descent—Sudden Transition from Arctic to Temperate Conditions—Delightful Surroundings—Weary Followers—Overtaken by Night—A Bivouac in the Forest—Fire in the Forest—Indian Sarcasm, 84-100

CHAPTER VII.

The Wild Character of Paradise Valley—Difficulties with Pack-Horses—A Remarkable Accident—Our Camp and Surroundings—Animal

Friends—Midsummer Flowers—Desolation Valley—Ascent of Hazel Peak—An Alpine Lake in a Basin of Ice—First Attempt to Scale Mount Temple—Our Camp by a Small Lake—A Wild and Stormy Night—An Impassable Barrier—A Scene of Utter Desolation—All Nature Sleeps—Difficulties of Ascent—The Highest Point yet Reached in Canada—Paradise Valley in Winter—Farewell to Lake Louise 101–118

CHAPTER VIII.

The Selkirks—Geographical Position of the Range—Good Cheer of the Glacier House—Charming Situation—Comparison between the Selkirks and Rockies—Early Mountain Ascents—Density of the Forest—Ascent of Eagle Peak—A Magnificent Panorama—A Descent in the Darkness—Account of a Terrible Experience on Eagle Peak—Trails through the Forest—Future Popularity of the Selkirks—The Forest Primeval—An Epitome of Human Life—Age of Trees—Forests Dependent on Humidity 119–136

CHAPTER IX.

Mount Assiniboine—Preparations for Visiting it—Camp at Heely's Creek—Crossing the Simpson Pass—Shoot a Pack-Horse—A Delightful Camp—A Difficult Snow Pass—Burnt Timber—Nature Sounds—Discovery of a Beautiful Lake—Inspiring View of Mount Assiniboine—Our Camp at the Base of the Mountain—Summer Snow-Storms—Inaccessibility of Mount Assiniboine . . . 137–157

CHAPTER X.

Evidence of Game—Discovery of a Mountain Goat—A Long Hunt—A Critical Moment—A Terrible Fall—An Unpleasant Experience—Habitat of the Mountain Goat—A Change of Weather—A Magnificent Panorama—Set out to Explore the Mountain—Intense Heat of a Forest Fire—Struggling with Burnt Timber—A Mountain Bivouac—Hope and Despair—Success at Last—Short Rations—Topography of Mount Assiniboine—The Vermilion River—A Wonderful Canyon—Fording the Bow River . . 158–182

Contents.

CHAPTER XI.

The Waputehk Range—Height of the Mountains—Vast Snow Fields and Glaciers—Journey up the Bow—Home of a Prospector—Causes and Frequency of Forest Fires—A Visit to the Lower Bow Lake—Muskegs—A Mountain Flooded with Ice—Delightful Scenes at the Upper Bow Lake—Beauty of the Shores—Lake Trout—The Great Bow Glacier 183-204

CHAPTER XII.

Sources of the Bow—The Little Fork Pass—Magnificence of the Scenery—Mount Murchison—Camp on the Divide—A High Mountain Ascent—Future of the Bow Lakes—Return down the Bow—Search for a Pass—Remarkable Agility of Pack-Horses—The "Bay" and the "Pinto"—Mountain Solitudes—Mount Hector—Difficult Nature of Johnston Creek—A Blinding Snow-Storm—Forty-Mile Creek—Mount Edith Pass . . 205-219

CHAPTER XIII.

HISTORICAL.

Origin and Rise of the Fur Trade—The Coureurs des Bois and the Voyageurs—Perils of the Canoe Voyages—The Hudson Bay Company and the Northwest Company—Intense Rivalry—Downfall of the Northwest Company—Sir Alexander Mackenzie—His Character and Physical Endowments—Cook's Explorations—Mackenzie Starts to Penetrate the Rockies—The Peace River—A Marvellous Escape—The Pacific Reached by Land—Perils of the Sea and of the Wilderness . . 220-236

CHAPTER XIV.

HISTORICAL.

Captain Cook's Explorations—The American Fur Company—First Exploration of the Fraser River—Expedition of Ross Cox—Cannibalism—Simplicity of a Voyageur—Sir George Simpson's Journey—Discovery

Contents.

of Gold in 1858—The Palliser Expedition—Dr. Hector's Adventures—Milton and Cheadle—Growth of the Dominion—Railroad Surveys—Construction of the Railroad—Historical Periods—Future Popularity of the Canadian Rockies . . . 237-257

CHAPTER XV.

The Pleasures of the Natural Sciences—Interior of the Earth—Thickness of the Crust—Origin and Cause of Mountains—Their Age and Slow Growth—System in Mountain Arrangement—The Cordilleran System—The Canadian Rockies—Comparison with Other Mountain Regions—Climate—Cause of Chinook Winds—Effect of High Latitude on Sun and Moon—Principal Game Animals—Nature of the Forests—Mountain Lakes—Camp Experiences—Effect on the Character . 258-275

INDEX . . 277-283

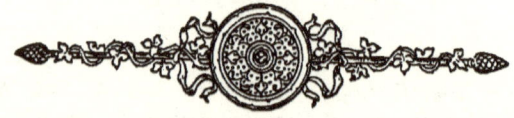

FULL-PAGE PHOTOGRAVURES.

	PAGE
MOUNT ASSINIBOINE . *Frontispiece*	
BANFF SPRINGS HOTEL	4
BOW RIVER AND CASCADE MOUNTAIN .	10
LAKE LOUISE .	18
MOUNT LEFROY AND MIRROR LAKE .	38
LAKE AGNES (In early July, 1895) .	42
TOM CHINIQUY (By courtesy of Mr. S. B. Thompson, New Westminster, B. C.) .	50
MOUNT TEMPLE, FROM THE SADDLE .	78
DISCOVERY OF PARADISE VALLEY	92
CAMP IN PARADISE VALLEY	108
MOUNT SIR DONALD, FROM EAGLE PEAK	126
HEAD OF ROCKY MOUNTAIN SHEEP . . .	132
NORTH LAKE .	152
SUMMIT LAKE, NEAR MOUNT ASSINIBOINE .	154
HEAD OF ROCKY MOUNTAIN GOAT (Shot July 18, 1895) . .	164
THE WAPUTEHK RANGE (Looking across the range from near Hector) .	184

xii   Full-Page Photogravures.

	PAGE
MOUNT DALY	192
UPPER BOW LAKE (Looking east)	196
UPPER BOW LAKE (Looking west)	200
SOURCE OF THE LITTLE FORK OF THE SASKATCHEWAN RIVER	206
STORM IN LITTLE FORK VALLEY	208
MOUNT HECTOR AND SLATE MOUNTAINS (From summit of a mountain near Little Fork Pass, 10,125 feet in altitude)	210
CAMP AT LITTLE FORK PASS	212
UPPER BOW LAKE (Looking south)	270
EMERALD LAKE AND MOUNT FIELD	272

ILLUSTRATIONS IN THE TEXT.

	PAGE
RUNDLE MOUNTAIN AND BOW RIVER	15
LAKE LOUISE (Looking toward chalet)	31
ANEMONES	40
A COOL RETREAT IN THE FOREST	75
SUMMIT OF MOUNT TEMPLE	115
GLACIER HOUSE	120
PEYTO	140
PACKING THE BUCKSKIN	142
CALYPSO	143
APPROACHING THE PASS	149
NORTH LAKE (Looking northwest)	157
HAUNT OF THE MOUNTAIN GOAT	165
MOUNT ASSINIBOINE (From northwest)	167
LAKE ON VERMILION PASS	181
READY TO MARCH	186
CAMP AT UPPER BOW LAKE	202
THE "BAY"	214
FALLS OF LEANCHOIL	249

CAMPING IN THE CANADIAN ROCKIES.

CHAPTER I.

Banff—Its Location—The Village—Tourists—Hotels—Topography of the Region—Rundle and Cascade Mountains—The Devil's Lake—Sir George Simpson's Journey to this Region—Peechee the Indian Guide—An Indian Legend—The Missionary Rundle—Dr. Hector—The Climate of Banff—A Summer Snow-Storm—The Mountains in Winter.

THE principal resort of tourists and sportsmen in the Rocky Mountains of Canada is Banff. The location of the town or village of Banff might be briefly described as being just within the eastern-most range of the Rocky Mountains, about one hundred and fifty miles north of the International boundary, or where the Canadian Pacific Railway begins to pierce the complex system of mountains which continue from this point westward to the Pacific coast.

Banff is likewise the central or focal point of the Canadian National Park. There is so much of scenic interest and natural beauty in the surrounding mountains and valleys, that an area of some two hundred and sixty square miles has been reserved in this region by the government and laid out with fine roads and bridle-paths to points of special interest. Order is enforced by a body of men known as the Northwest Mounted Police, a detachment of which is stationed at Banff. This organization has been wonderfully effective for many years past in preserving the authority of the laws throughout the vast extent of northwestern Canada by means of a number of men that seems altogether insufficient for that purpose.

The small and scattered village of Banff occupies a flat plain near the Bow River. This large stream, the south branch of the Saskatchewan, one of the greatest rivers of North America, is at this point not only deep and swift but fully one hundred yards in width. A fine iron bridge spans the river and leads to the various hotels all of which are south of the village. The permanent population numbers some half thousand, while the various stores, dwellings, and churches have a general air of neatness and by their new appearance suggest the fact that the history of Banff extends back only one decade.

During the summer season, the permanent population of Banff is sometimes nearly doubled by a great invasion of tourists and travellers from far distant regions. Overland tourists from India, China, Ceylon, and England, the various countries of Europe and the Dominion of Canada, but chiefly from the United States, form the greater part of this cosmopolitan assemblage, in which, however, almost every part of the globe is occasionally represented. Some are bent on sport with rod or gun ; others on mountaineering or camping expeditions, but the great majority are en route to distant countries and make Banff a stopping-place for a short period.

Arrived at Banff, the traveller is confronted by a line of hack drivers and hotel employes shouting in loud voices the names and praises of their various hotels. Such sights and sounds are a blessed relief to the tourist, who for several days has witnessed nothing but the boundless plains and scanty population of northwestern Canada. The chorus of rival voices seems almost a welcome back to civilization, and reminds one in a mild degree of some railroad station in a great metropolis. On the contrary, the new arrival finds, as he is whirled rapidly toward his hotel in the coach, that he is in a mere country village surrounded on all sides by high mountains, with here and there patches of perpetual snow near their lofty summits.

Though the surrounding region, the adjacent mountains, and valleys represent nature in a wild and almost primitive state, one may remain at Banff attended by all the comforts of civilization. The several hotels occupy more or less scattered points in the valley south from the village. The one built and managed by the railroad stands apart from the village on an eminence overlooking the Bow River. It is a magnificent structure capable of accommodating a large number of guests. From the verandas and porches one may obtain a fine panoramic view of the surrounding mountains, and on the side towards the river the view combines water, forest, and mountain scenery in a most pleasing manner. The Bow River, some three hundred feet below, comes in from the left and dashes in a snowy cascade through a rocky gorge, then, sweeping away towards the east, is joined by the Spray River, a mad mountain torrent deep and swift, but clear as crystal, and with cold water of that deep blue color indicating its mountain origin. The wonderful rapidity with which these mountain streams flow is a source of astonishment and wonder to those familiar only with the sluggish rivers of lowland regions. Standing on the little iron bridge which carries the road across the stream and looking down on the water, I have often imagined I was at the stern of an ocean greyhound, so rapidly does each ripple or inequality sweep under and away from the eye. Though the water is less than a yard in depth, the current moves under the bridge at the rate of from nine to ten miles an hour.

The best point from which to get a good general idea of the topography of Banff and its surroundings is from the summit of a little hill known as Tunnel Mountain. It is centrally located in the wide valley of the Bow, above which it rises exactly 1000 feet, an altitude great enough to make it appear a high mountain were it not dwarfed by its mighty neighbors. The view from the summit is not of exceeding grandeur, but is well worth the labor of the climb, especially as a good path, with occasional seats for the weary, makes the walk an easy one. The top of the mountain is still far below the tree line, though the earth is too thin to nourish a rich forest. The soil was

all carried away in the Ice Age, for there are abundant proofs that this mountain was once flooded by a glacier coming down the Bow valley. The bare limestone of the summit is grooved in great channels pointing straight up the Bow valley. In some places scratches made by the ice are visible, and there are many quartz boulders strewed about which have been carried here from some distant region.

The meandering course of the Bow River, the village, the hay meadows and grassy swamps, all form a pretty picture in the flat valley below. The eastern face of Tunnel Mountain is wellnigh perpendicular. The trail leads along near the summit and allows thrilling views down the sheer precipice to the flat valley of the Bow River far below. The trees and prominent objects of the landscape seem like toys, and the adjacent plains resemble a colored map. There are no houses or dwellings in view on this side, but a drove of horses grazing contentedly in a pasture near the river, awaiting their turn to be sent out into the mountains in the pack train of some sportsman or mountaineer, gives life and animation to the scene. On either side are two high mountains, conspicuous by their unusual outlines and great altitude. The one to the south is Rundle Mountain. It rises in a great curving slope on its west side, and terminates in a rugged escarpment with precipitous cliffs to the east, which tower in wonderful grandeur more than 5000 feet above the flood plains of the Bow River near its base.

On the opposite side is Cascade Mountain, which is remarkable in being of almost identical height, and is in fact just two feet lower, as determined by the topographical survey. The name of this mountain was given by reason of a large stream which falls from ledge to ledge down the cliffs of its eastern face in a beautiful cascade. Both this and Rundle Mountain are composed of the old Devonian and Carboniferous limestones, the strata of which are plainly visible. The structure is that of a great arch or anticline which has been completely overturned, so that the older beds are above the newer. Several miles towards the east, the end of Devil's Lake may be seen appearing through a notch in the mountains. A fine road nine miles in length has been made to this lake and is one of the most popular drives in the vicinity of Banff. The lake is very long and narrow, about nine miles in length by three fourths of a mile in extreme breadth. The scenery is grand, but rather desolate, as the bare mountain walls on either side of the lake are not relieved by forests or abundant vegetation of any kind. The lake is, however, a great resort for sportsmen as it abounds in large trout, of which one taken last year weighed thirty-four pounds. The name of the lake gives illustration of the tendency among savages and civilized people to dedicate prominent objects of nature to the infernal regions or the master spirit thereof. There is no apparent limit to the number of places named after the Devil and his realm, while the names suggested by more congenial places are conspicuous by their absence. The original name, Lake Peechee, was given by Sir George Simpson in honor of his guide.

The scattered threads of history which relate to this part of the Rocky Mountains are suggested by these names and indeed this lake has an unusual interest for this reason. In a region where explorations have been very few and far between, and where only the vague traditions of warlike events among the Indians form a great part of the history, each fragment and detail set forth by the old explorers acquires an increased interest.

Previous to the arrival of the railroad surveyors, the chief men on whom our attention centres are Sir George Simpson, Mr. Rundle, and Dr. Hector.

The expedition of Sir George Simpson possesses much of interest in every way. He claims to have been the first man to accomplish an overland journey around the world from east to west. After having traversed the greater part of the continent of North America, he entered the stupendous gates of the Rocky Mountains in the autumn of 1841. He travelled with wonderful rapidity, and was wont to cover from twenty to sixty miles a day, according to the nature of the country. His outfit consisted of a large band of horses, about forty-five in number, attended by cooks and packers sufficient for the needs of this great expedition. Nevertheless the long cavalcade of animals, when spread out in Indian file along the narrow trails were difficult to manage, and it not infrequently happened that on reaching camp several horses proved to be missing, a fact which would necessitate some of the men returning fifteen or twenty miles in search of them.

Passing to the south of the Devil's Head, a remarkable and conspicuous mountain which may be recognized far out on the plains, Sir George Simpson entered the valley occupied by the lake. In this part of his journey he was guided by a half-breed Indian named Peechee, a chief of the Mountain Crees. Peechee lived with his wife and family on the borders of this lake, and Simpson named it after him, a name, however, which never gained currency. Dr. Dawson transferred the name to a high mountain south of the lake, and substituted the Indian name of Minnewanka, or in English, Devil's Lake.

The guide Peechee seems to have possessed much influence among his fellows, and whenever, as was often the case, the Indians gathered around their camp-fires and gossiped about their adventures, Peechee was listened to with the closest attention on the part of all. Nothing more delights the Indians than to indulge their passion for idle talk when assembled together, especially when under the soothing and peaceful influence of tobacco,—a fact that seems strange indeed to those who see them only among strangers, where they are wont to be remarkably silent.

A circumstance of Indian history connected with the east end of the lake is mentioned by Sir George Simpson, and admirably illustrates the nature of savage warfare. A Cree and his wife, a short time previously, had been tracked and pursued by five Indians of a hostile tribe into the mountains to a point near the lake. At length they were espied and attacked by their pursuers. Terrified by the fear of almost certain death, the Cree advised his wife to submit without defending herself. She, however, was possessed of a more courageous spirit, and replied that as they were young and had but one life to lose they had better put forth every effort in self-defence. Accordingly she raised her rifle and brought down the foremost warrior with a well aimed shot. Her husband was now impelled by desperation and shame to join the contest, and mortally wounded two of the advancing foe with arrows. There were now but two on each side. The fourth warrior had, however, by this time reached the Cree's wife and with upraised tomahawk was on the point of cleaving her head, when his foot caught in some inequality of the ground and he fell prostrate. With lightning stroke the undaunted woman buried her dagger in his side. Dismayed by this unexpected slaughter of his companions, the fifth Indian took to flight after wounding the Cree in his arm.

Rundle Mountain, which has been already mentioned and which forms one of the most striking mountains in the vicinity of Banff, is named after a Wesleyan missionary who for many years carrried on his pious labors among the Indians in the vicinity of Edmonton. Mr. Rundle once visited this region and remained camped for a considerable time near the base of Cascade Mountain, probably shortly after Sir George Simpson explored this region. The work of Mr. Rundle among the Indians appears to have been highly successful, if one may judge by the present condition of the Stoneys, who are honest, truthful, and but little given to the vices of civilization.

Even to this day the visitor may see them at Banff dressed in partly civilized, partly savage attire, or on rare occasions decked out in all the feathers and beaded belts and moccasins that go to make up the sum total of savage splendor.

Our attention comes at last to Dr. Hector, who was connected with the Palliser expedition. It is exceedingly unfortunate that the blue-book in which the vast amount of useful information and interesting adventure connected with this expedition is so clearly set forth should be now almost out of print. There are no available copies in the United States or Canada and but very few otherwise accessible. Dr. Hector followed up the Bow River and passed the region now occupied by Banff in the year 1858. He was accompanied by the persevering and ever popular botanist, Bourgeau. Under the magic spell of close observation and clear description, the most commonplace affairs assume an unusual interest in all of Dr. Hector's reports. It is very evident that game was much more abundant in those early days than at present. For instance, Dr. Hector's men shot two mountain sheep near the falls of the Bow River, which are but a few minutes' walk from the hotel. Likewise when making a partial ascent of the Cascade Mountain, Dr. Hector came on a large herd of these noble animals, concerning which so many fabulous tales of their daring leaps down awful precipices have been told. He also mentions an interesting fact about the death of a mountain goat. An Indian had shot a goat when far up on the slope of

The Climate of Banff.

Cascade Mountain, but the animal, though badly wounded, managed to work its way around to some inaccessible cliffs near the cascade. Here the poor animal lingered for seven days with no less than five bullets in its body, till at length death came and it fell headlong down the precipice.

The climate of Banff during the months of July and August is almost perfection. The high altitude of 4500 feet above the sea-level renders the nights invariably cool and pleasant, while the mid-day heat rarely reaches 80° in the shade. There is but little rain during this period and in fact there are but two drawbacks,—mosquitoes and forest-fire smoke. The mosquitoes, however, are only troublesome in the deep woods or by the swampy tracts near the river. The smoke from forest fires frequently becomes so thick as to obscure the mountains and veil them in a yellow pall through which the sun shines with a weird light.

An effect of the high northern latitude of this part of the Rocky Mountains is to make the summer days very long. In June and early July the sun does not set till nine o'clock, and the twilight is so bright that fine print can be read out doors till eleven o'clock, and in fact there is more or less light at midnight.

In June and September one never knows what to expect in the way of weather. I shall give two examples which will set forth the possibilities of these months, though one must not imagine that they illustrate the ordinary course of events. In the summer of 1895, after having suffered from a long period of intensely hot weather in the east, I arrived at Banff on the 14th of June. It was snowing and the station platform was covered to a depth of six inches. The next day, however, I ascended Tunnel Mountain and found a most extraordinary combination of summer and winter effects. The snow still remained ten or twelve inches deep on the mountain sides, though it had already nearly disappeared in the valley. Under this wintry mantle were many varieties of beautiful flowers in full bloom, and, most conspicuous of all, wild roses in profusion, apparently uninjured by this unusually late snow-storm. I made a sad discovery near the top of the mountain. Seeing a little bird fly up from the ground apparently out from the snow, I examined more closely and observed a narrow snow-tunnel leading down to the ground. Removing the snow I found a nest containing four or five young birds all dead, their feeble spark of life chilled away by the damp snow, while the mother bird had been, even when I arrived, vainly trying to nurse them back to life.

This storm was said to be very unusual for the time of year. The poplar trees in full summer foliage suffered severely and were bent down to the ground in great arches, from which position they did not fully recover all summer, while the leaves were blighted by the frost. As a general rule, however, mountain trees and herbs possess an unusual vitality, and endure snow and frost or prolonged dry weather in a remarkable manner. The various flowers which were buried for a week by this late storm appeared bright and vigorous after a few warm days had removed the snow.

Toward the end of September, 1895, there were two or three days of exceptionally cold weather, the thermometer recording 6° Fahrenheit one morning. I made an ascent of Sulphur Mountain, a ridge rising about 3,000 feet above the valley, on the coldest day of that period. The sun shone out of a sky of the clearest blue without a single cloud except a few scattered wisps of cirrus here and there. The mountain summit is covered with a few straggling spruces which maintain a bare existence at this altitude. The whole summit of the mountain, the trees, and rocks were covered by a thick mantle of snow, dry and powdery by reason of the severe cold. The chill of the previous night had condensed a beautiful frost over the surface of the snow everywhere. Shining scales of transparent ice, thin as mica and some half-inch across, stood on edge at all possible angles and reflected the bright sunlight from thousands of brilliant surfaces. This little glimpse of winter was even more pleasing than the view from the summit, for the mountains near Banff do not afford the mountain climber grand panoramas or striking scenery. They tend to run in long regular ridges, uncrowned by glaciers or extensive snowfields.

A never failing source of amusement to the residents of Banff, as well as to those more experienced in mountain climbing, is afforded by those lately arrived but ambitious tourists who look up at the mountains as though they were little hills, and proceed forthwith to scale the very highest peak on the day of their arrival. A few years ago some gentlemen became possessed of a desire to ascend Cascade Mountain and set off with the intention of returning the next day at noon. Instead of following the advice of those who knew the best route, they would have it that a course over Stoney Squaw Mountain, an intervening high ridge, was far better. They returned three days later, after having wandered about in burnt timber so long that, begrimed with charcoal, they could not be recognized as white men. It is not known whether they ever so much as reached the base of Cascade Mountain, but it is certain that they retired to bed upon arriving at the hotel and remained there the greater part of the ensuing week.

Cascade Mountain, however, is a difficult mountain to ascend, not because there are steep cliffs or rough places to overcome, but because almost every one takes the wrong slope. This leads to a lofty escarpment, and just when the mountaineer hopes to find himself on the summit, the real mountain appears beyond, while a great gulf separates the two peaks and removes the possibility of making the ascent that day.

Banff, with its fine drives and beautiful scenery, its luxurious hotels and delightful climate, will ever enjoy popularity among tourists. The river above the falls is wide and deep and flows with such gentle current as to render boating safe and delightful. The Vermilion lakes, with their low reedy shores and swarming wild fowl, offer

Vermilion Lakes. 15

charming places for the canoe and oarsman, at least when the mosquitoes, the great pest of our western plains and mountains, temporarily disappear. Nevertheless, the climate of Banff partakes of the somewhat dryer nature of the lesser and more eastern sub-ranges of the Rocky Mountains. There is not sufficient moisture to nourish the rich forests, vast snow-fields, and thundering glaciers of the higher ranges to the west, which in imagination we shall visit in the ensuing chapters.

RUNDLE MOUNTAIN AND BOW RIVER.

CHAPTER II.

*Lake Louise—First Impressions—An Abode of Perpetual Winter—
The Chalet—Visitors—Stirring Tales of Adventure—Primeval Forests—
Forest Fires—Mosquitoes and Bull-Dog Flies—Mortal Combats between
Wasps and Bull-Dogs—The Old Chalet—Morning on the Lake—Approach
of a Storm—Sublimity of a Mountain Thunder-Storm—Cloud Effects—
The Lake in October—A Magnificent Avalanche from Mt. Lefroy—A
Warning of Approaching Winter.*

LAKE LOUISE is one of the most beautiful sheets of water in the Canadian Rockies. Many who have travelled extensively say it is the most charming spot they have ever beheld. The lake is small, but there is a harmonious blending of grandeur and quiet beauty in the surrounding mountains which in some way makes a perfect picture out of lofty snow peaks in the distance and dark forested slopes near at hand.

The lake is a little more than a mile long and about one fourth of a mile wide. The outline is remarkably like that of the left human foot. Forests come down nearly to the water's edge on all sides of the lake, but there is a narrow margin of rough angular stones where the ripples from the lake have washed out the soil and even undermined the trees in some places. The water is a blue-green color, so clear that the stones on the bottom and the old water-logged trunks of trees, long since wrested from the shores by storms and avalanches, may be discerned even in several fathoms of water. The lake is 230 feet deep in the centre, and the bottom slopes down very suddenly from the shores.

The west shore makes a gently sinuous or wavy line, forming little bays and capes. Ever new and artistic foregrounds are thus presented, with the forest making a retreating line of vegetation down the shore. Nothing could be more beautiful than this border of the lake, rough and tangled though it is, with a strange mingling of sharp boulders and prostrate trees covered with moss and half concealed by copses of alder bushes and flowering shrubs.

I shall never forget my first view of Lake Louise. From the station, the old trail, constantly ascending as it approaches the lake, leads its irregular course through the forest. After a walk of nearly three miles, partial glimpses of the lake and surrounding mountains were obtained from among the tall spruce trees. A short rapid descent of a small ridge placed us on the borders of the lake.

It would be difficult indeed to give even a partial description of the scene. Imagine a cool morning with the rising sun just beginning to touch the surface of a mountain lake. The air is tranquil and calm so that the glassy surface of the water mirrors the sky and mountains perfectly. In the realm of sound, too, all is repose but for the call of birds near at hand among the balsam trees. From the shores of the lake on either side rise great mountains, showing cliffs and rocky ledges or long sweeping slopes of forest to the tree line. Higher still are bare slopes, crags, ledges, and scattered areas of snow. At the end of the lake a great notch in the nearer mountains reveals at a distance the wall-like, lofty mass of Mount Lefroy. This most imposing snowy mountain stands square across the gap, and with a sharp serrated cliff piercing the very vault of heaven, shuts off the view and forms the most conspicuous object of all. The lower part of the mountain is a vertical cliff or precipice where the longitudinal strata are distinctly visible. Above, rise alternating slopes covered with perpetual snow and hanging glaciers, the white-blue ice of which is splintered by deep rents and dark yawning crevasses. This mountain forms part of the continental water-shed, for on the other side the melting snows finally reach the Pacific Ocean, while on the near side the snows swept into the valleys by avalanches, and melted by the warmer air of lower altitudes, find their way at length into the Saskatchewan River and Hudson Bay.

There is something wonderfully attractive about this mountain. The pleasure grows as one continues to gaze at the immense mass ; harsh and stern and cold though it be, it excites awe and wonder as though here were the rocky foundation and substratum of the globe. This is the abode of perpetual winter, where ice and snow and

bleak rocks exist apart. Here all is grand but menacing, dangerous, and forbidding. And these high mountains and deep valleys, suggesting that some awful storm at sea had become petrified into colossal waves to stand at rest forever, have been carved out by rain and running water, frost and change of temperature, through the lapse of countless ages.

Our attention finally came to the quiet beauty of the surrounding vegetation, where among the scattered skirmishers of the forest are flowering shrubs, and in the more open grassy places forming the swampy borders of the lake, are many bright flowers. The white mountain anemones in several varieties, the familiar violets, the yellow columbine with beautiful pendent blossoms claiming relationship to its Eastern cousin with scarlet flowers, the fragrant spiranthes, and orchids with pale-green flowers, resembling insects on a leafy stem, may all be seen in profusion near the north side of the lake. These humble herbs, with their gaudy coloring, are the growth of a single season, but on all sides are copses of bushy plants which endure the long winter, some of them clad in a garb of evergreen and, like the annual plants, bearing elegant floral creations. The most conspicuous is the sheep laurel, a small bush adorned with a profusion of crimson-red flowers, each saucer-shaped, hanging in corymbs among the small green leaves. Various shrubs with white flowers, some small and numerous, others large and scattered, make a contrast to the ever present laurel, while the most beautiful of all is a species of mountain rhododendron, a large bush, the most elegant among the mountain heaths, with large white flowers in clustered umbels. In early July this bush may be found, here and there, scattered sparingly in the forest in full blossom at the level of Lake Louise, but after this one must seek ever higher on the mountain side as the advancing summer creeps to altitudes where spring is later.

The early morning visitor turns with sharpened appetite to the hotel, if we may call it such,—a little Swiss chalet of picturesque architecture built on an eminence in full view of the lake. Here the tourist may live in rustic comfort for a day, or for weeks, should he desire to prolong his visit.

Tourists come sparingly to Lake Louise. Unlike Banff with its varied attractions, there is little here outside of nature, and few have the power to appreciate nature alone. Of those who do come, only a small number really see the lake with its forests and mountains combined in exquisite attractiveness. They see the outlines of mountains, but know not whether they are near or distant, nor whether their scale is measured in yards or miles; they see the water of the lake, but not the reflections in it, the ever changing effects of light and shade, sun and shadow, ripple and calm. There are trees tall and slender, but whether they be spruce or pine, larch or hemlock, is all the same; and as to the flowers—some are differently colored from others.

A visitor to the lake once asked in good faith, apparently, if the mountains at the head of the lake were not white from chalk; another, why the water of the stream

—which leads out from the lake and rushes in roaring cascades over its rocky channel toward the Bow River— runs so fast down hill.

Fortunately, however, those who are not blessed with that ever present source of pleasure, a love for nature, at least to a slight degree, are exceptional. Nevertheless, that most people lose much pleasure from a lack of close observation is often painfully evident. I have seen, altogether, several hundred tourists arrive at the lake, coming as they do in small parties, or singly, from day to day, and have found it a very interesting study to observe their first impressions as the lake bursts on their view. Some remain motionless studying the details of the scene, usually devoting their chief attention to the lake and forests, but less to the mountains, for mountains are the least appreciated of all the wonders of nature, and are not fully revealed except after years of experience. Others glance briefly and superficially towards the lake, and rush hastily into the chalet for breakfast, balancing their love for nature against hunger for material things in uneven scale. Some remain a week or ten days, but the great majority spend a single day and leave, feeling that they have exhausted the charms of the place in so short a time. A single day amid surroundings where there are such infinite possibilities of change in cloud and storm, heat and cold, the dazzling glare of noon, or the calm romantic light of a full moon, and the slow progress of the seasons, gives but one picture, a single mood from out a thousand, and it may perchance be the very worst of all.

Upon climbing the steps to the open porch of the chalet and entering the large spacious sitting-room, the eye falls at once on a fireplace of old-time proportions, and within its walls of brick, huge logs are burning, with more vigor indeed but hardly less constancy than the ancient fires of the Vestal Virgins. Round this spacious hearth visitors and guests gather, for the air at Lake Louise is always sharp at morning and evening. Indeed, frosts are not rare throughout the summer and may occur any week even in July and August. The high altitude of the lake, which is a little more than 5600 feet above sea-level, is in great part the cause of this bracing weather. On the hottest day that I have ever seen at the lake in the course of three summers the thermometer registered only 78°.

The visitors who come to Lake Louise are of the same cosmopolitan character and varied nationality as those at Banff. Often of a cold night have I sat by the large fire, our only source of light, and listened to tales of adventure told by those who have visited the most distant and unfrequented parts of the earth. Englishmen, who have spent the best years of their life in India, were among our entertainers, and while beverages varying in nature according to nationality or tastes of each were passed around, I have heard thrilling accounts of leopard and tiger hunts in the jungle, blood-curdling tales of treachery and massacre or daring exploits in the Indian wars, and rare experiences in unknown parts of Cashmere and Thibet.

Primeval Forests.

Though the great majority of visitors to the lake are strangers, there are some half-dozen whose familiar faces reappear each successive season ; like pilgrims they make this region the termination of a long annual journey, and here worship in "temples not built by human hands." Among these lovers of nature, far distant England and Ceylon are represented no less than the nearer cities of the United States. The peculiar charms of this locality present an inexhaustible treasurehouse of delightful experiences that grow by familiarity. One's impressions of the beauty of the lake increase year by year as the full meaning of each detail becomes more thoroughly appreciated.

A fact of great importance, which goes far to make up the *ensemble* of the surroundings of Lake Louise, is the perfect condition of the forests, which rise in uniform, swelling slopes of dark-green verdure from the rocky shores of the lake far up the mountain sides to those high altitudes where the cold air suggests an eternal winter and dwarfs the struggling trees into mere bushes. The frequent forest fires, which have wrought so much destruction throughout the entire Canadian Rockies, have not as yet swept through this valley. The great spruces and balsams of this primeval forest indicate by their size that for hundreds of years no fire has been through this region. Some large tree stumps near the chalet show hundreds of rings, and one that I counted started to grow in the year 1492, when Columbus set forth to discover the western world.

Nevertheless, on hot days after a long period of dry weather, when the air is laden with the fragrant odor of the dripping balsam and of the dry resin hardened in yellow tears on the scarred trunks of the trees, and when the dead lower branches hung with long gray moss seem to offer all the most combustible materials, one feels certain that the slightest spark would result in a terrible conflagration. Apparently, however, the past history of this valley has never recorded a fire, whether started by careless Indian hunters or that frequent cause, lightning. So far as I am aware, there are no layers of buried charcoal or reddened soil under the present forest which would indicate an ancient fire.

Some years ago—apparently more than twenty,—a fire destroyed the forest near the station of Laggan, which is less than two miles from the lake in a straight line. The fire approached within a mile of the lake and then died out. There are two causes which will always tend to preserve these beautiful forests if the visitors are not careless and counteract them. The prevalent wind is out of the valley toward the Bow valley, so that a fire would naturally be swept away from the lake. Another cause is the natural moisture of this upland region. The very luxuriance of the vegetation indicates this, while in the early morning the whole forest often seems reeking with moisture, even when there has been no rain for weeks. The chill of night appears to condense a heavy dew under the trees and moistens all the vegetation, so that the forest rarely becomes so exceedingly dry as often happens in wide valleys at lower altitudes.

Mosquitoes and Bull-Dogs.

Though the scenery and climate at Lake Louise seem almost ideally perfect during the summer time, nature always renders compensation in some form or other, and never allows her creatures to enjoy complete happiness. The borders of the lake and the damp woods breed myriads of mosquitoes, which conspire to annoy and torture both man and beast. They appear early in spring and suddenly vanish about the 15th or 20th of August each year. The chill of night causes them to disappear about ten o'clock in the evening, not to be seen again until the atmosphere begins to grow warm in the morning sun.

Another insect pest is a species of fly called the "bull-dog," a name suggested by its ferocious bite. These large insects are about an inch in length and are armed with a formidable set of saws with which they can rapidly cut a considerable hole through the skin of a man or the hide of a horse. The bull-dogs frequent the valleys of the Canadian Rockies, varying locally in their numbers, and seem to prefer low altitudes and a considerable degree of heat, for they are always most voracious and numerous on hot dry days. These flies, when numerous, will almost make a horse frantic. Their bite feels like a fiery cinder slowly burning through the skin, but fortunately they do not cause much trouble to man, for they are led by instinct to seek the rough surfaces of animals and almost invariably light on the clothes instead of the hands or face. They have a most blood-thirsty and cruel enemy in the wasp, and if it were not for the inexhaustible supply of the bull-dogs, the wasps would annihilate the species. Nothing in the habits of insects could be more interesting than the strange manner in which the wasps set out deliberately in pursuit of a bull-dog fly, to overtake and seize the clumsy victim in mid air. Both insects fall to the ground with a terrible buzzing and much circling about while the mad contest goes on. Meanwhile the wasp works with the rapidity of lightning, and with its sharp powerful jaws dissevers legs and wings, which fall scattered in the melee, till the bull-dog is rendered helpless and immovable. Last of all, the wasp cuts off the head of its victim, then leaves the lifeless and limbless body in order to continue the chase.

I have seen a wasp thus dismember and kill one of these large flies in less than thirty seconds. They seem to perform their murderous acts out of pure pleasure, as they do not linger over their prey after the victim is dead.

The water of Lake Louise is too cold to admit of bathing except in a very brief manner. The temperature of the water near the first of August is about 56°.

The old chalet, built in rustic fashion with unhewn logs, was placed near the lake shore much closer than the present building. One day in 1893, when every one was absent, the building caught fire and burned to the ground. Remarkably enough the forest did not take fire, though some of the trees were close to the building.

Usually in the early morning, before the sun has warmed the atmosphere and started the breezes of daytime into motion, the lake is tranquil and its surface resembles a great mirror. About nine o'clock, the first puffs of wind begin to make little cat's-paws at the far end of the lake, which widen and extend until finally the whole water becomes rippled. A gentle breeze continues to sweep down the lake from the snow mountains toward the Bow valley all day long, and the water rarely becomes smooth till after sunset. This is the usual order of events in fair weather, a condition which may continue for several weeks without a drop of rain.

The approach and progress of a storm, the wonderful atmospheric changes attending it, and the ever moving clouds obscuring the mountain tops reveal the lake in the full grandeur of its surroundings. An approaching storm is first announced by scattered wisps of cirrus cloud, which move slowly and steadily from the west in an otherwise blue sky. In the course of twenty-four hours the cirrus clouds have become so thick that they often resemble a thin haze far above the highest mountains. The sun with paled light can no longer pierce this ever thickening hazy veil. The wind blows soft and warm from out the south or southwest, and generally brings up the smoke of forest fires from the Pacific coast, and renders the atmosphere still more obscure, till at length the sun appears like a great ball of brass set in a coppery sky. The trees and grass appear to change their color and assume a strange vivid shade of green in the weird light. Sometimes light feathery ashes are wafted over the high mountains south of the lake and settle down gently like flakes of snow. The falling barometer announces the coming storm, and presently another layer of clouds, the low-lying cumulus, form just above the highest peaks and settle gradually lower till they touch the mountain tops. Rain soon follows, the clouds settle till they almost rest on the water of the lake, and the wind increases in violence.

Sometimes thunder-storms of considerable fury sweep through the valley and among the mountains, one after another for several days. A violent thunder-storm at night among these lofty mountains is one of the grandest phenomena of nature. The battling of the elements, the unceasing roar of the wind in the forest, and the crash of thunder redoubled by echoes from the rocky cliffs,—all conspire to fill the imagination with a terrible picture of the majesty and sublimity of nature. From the lake there comes up a low, hoarse murmur, not the roar of ocean surf, but the lesser voice of a small mountain lake lashed to fury and beating with its small waves on a rocky shore. The noise of the forest, the sound of colliding branches as the tall trees sway to and fro in the furious wind, and the frequent crack and crash of dead forest giants overcome by the elements form the dull but fearful monotone, above which the loud rumble of thunder rises in awful grandeur. These are the sounds of a mountain storm.

The bright flashes of lightning reveal a companion picture, for in the momentary light succeeded by absolute darkness the lake is revealed covered with foamy white caps. The forests on the mountain side seem to yield to the blast like a field of wheat in a summer breeze, and the circling clouds sweep about the mountain slopes and conceal all but their bases.

Should the storm clear away during the daytime one

may witness grand cloud effects. The low-hanging masses of clouds left behind by the battling elements slowly rise and occasionally reveal small areas of blue sky among the moving vapors. Gentle puffs of air sweep over the calm surface of the water, making little areas of ripples here and there, only to be succeeded by a tranquil calm, as if the storm spirit were sending forth his dying gasps intermittently. While the air is thus calm below, the circling wisps of vapor high up on the mountain, rising and descending, show that the battle between the sun and the clouds is still raging. From above the saturated forests, the rising vapors condense and increase in size till at length, caught in some counter-current, they are swept away or carried downward, while the dissolving cloud spreads out in wisps and streamers till suddenly it disappears into transparent air,—a veritable cloud ghost. At length the mountain tops appear once more, white in a light covering of new snow, and, as the great masses of cumulus rise and disappear the sky appears of that deep blue-black color peculiar to mountain altitudes, while the sun shines out with dazzling brilliancy through the clear atmosphere.

The last visit I made to Lake Louise was toward the middle of October, 1895. A very snowy, disagreeable September had been followed by a long period of milder weather with much bright sunshine. The new snow, which had been quite deep near the lake, had altogether disappeared except high up on the mountain side. It was the true Indian summer, a season with a certain mellow charm peculiar to it alone, characterized by clear sunny weather, a calm atmosphere, a low, riding sun, and short days. Most of the flowers were withered. The deciduous bushes, lately brilliant from frost, were rapidly losing their foliage, and the larches were decked in pale yellow, far up near the tree line. However, the greater part of the vegetation is evergreen, and the spruces, balsams, and pines, the heaths, ericaceous plants, and the mosses contrive to set winter at nought by wearing the garb of a perpetual summer in a region where snow covers the ground three fourths of the year.

I could not resist the temptation as the morning train rolled up to the station at Laggan to get off for the day and make another visit to the lake. The sunrise had been unusually brilliant and there was every promise of a fine day. There is rarely much color at sunrise or sunset in the mountains. The dry clear atmosphere has little power to break up the white light into rainbow colors and give the brilliancy of coloring to be seen near the sea-coast or in the lowlands. The tints are like the air itself—pure, cold, and clear. With more truth they might be called delicate shades or color suggestions. They recall those exquisite but faint hues seen in topaz or tourmaline crystals, or transparent quartz crystals, wherein the minutest trace of some foreign mineral has developed rare spectrum colors and imprisoned them forever. Oftimes the snow of the mountain tops is thus tinted a bright clear pink, beautifully contrasted against the intensely blue sky. I have never seen a deep red on the mountains or clouds at these altitudes. The effect

The Lake in October.

of forest-fire smoke is to give muddy colors: the sun resembles a brazen globe, and the sky becomes coppery in appearance.

After breakfast at the station house, I set off over the hard frozen road toward the lake. I carried my camera and luncheon on my back, my only companion being a small dog which appeared ready for exercise. The air was frosty and cold; the low-riding sun had not as yet struck into the forest trees and removed the rime from the moss and leaves on the ground.

In somewhat less than an hour, I arrived at the lake. All was deserted; the chalet closed, the keeper gone, and the tents taken down. Even the boats, which usually rested near the shore, had been put under cover. The cold air was perfectly calm, and my vapory breath rose straight upwards. The mirror surface of the water was disturbed by some wild fowl—black ducks and divers—which swarm on the lake

LAKE LOUISE LOOKING TOWARD CHALET.

at this season. Their splashings, and the harsh cries of the divers came faintly over the water. It seemed strange that these familiar haunts could appear so fearfully wild and lonely merely because man had resigned his claim to the place and nature now ruled alone. All at once a wild unearthly wail from across the water, the cry of a loon, one of the most melancholy of all sounds, startled me, and gave warning that activity alone could counteract the effect of the imagination.

Accordingly I walked down the right shore of the lake with the intention of going several miles up the valley and taking some photographs of Mount Lefroy. The flat bushy meadows near the upper end of the lake were cold, and all the plants and reedy grass were white with the morning frost. The towering cliffs and castle-like battlements of the mountains on the south side of the valley shut out the sun, and promised to prevent its genial rays from warming this spot till late in the afternoon, if at all, for a period of several months. In the frozen ground, as I followed the trail, I saw the tracks of a bear, made probably the day before. Bruin had gone up the valley somewhere and had not returned as yet, so there was a possibility of making his acquaintance.

I was well repaid for my visit this day, as a magnificent avalanche fell from Mount Lefroy. Mount Lefroy is a rock mountain rising in vertical cliffs from between two branches of a glacier which sweep round its base. A hanging glacier rests on the highest slope of the mountain, and, ascending some distance, forms a vertical face of

A Magnificent Avalanche.

ice nearly three hundred feet thick at the top of a great precipice. The highest ridge of the mountain is covered with an overhanging cornice of snow, which the storm winds from the west have built out till it appears to reach full one hundred feet over the glacier below. At times, masses of ice break off from the hanging glacier and fall with thundering crashes to the valley far below.

I was standing at a point some two miles distant looking at this imposing mountain, when from the vertical ice wall a great fragment of the glacier, some three hundred feet thick and several times as long, broke away, and, slowly turning in mid-air, began to fall through the airy abyss. In a few seconds, amid continued silence, for the sound had not yet reached me, the great mass struck a projecting ledge of rock after a fall of some half thousand feet, and at the shock, as though by some inward explosion, the block was shivered into thousands of smaller fragments and clouds of white powdery ice. Simultaneously came the first thunder of the avalanche. The larger pieces led the way, some whirling around in mid-air, others gliding downward like meteors with long trains of snowy ice dust trailing behind. The finer powdered debris followed after, in a long succession of white streamers and curtains resembling cascades and waterfalls. The loud crash at the first great shock now developed into a prolonged thunder wherein were countless lesser sounds of the smaller pieces of ice. It was like the sound of a great battle in which the sharp crack of rifles mingles with the roar of artillery. Leaping from ledge to ledge with ever increasing velocity, the larger fragments at length reached the bottom of the precipice, while now a long white train extended nearly the whole height of the grand mountain wall 2500 feet from base to top.

Imagine a precipice sixteen times higher than Niagara, nearly perpendicular, and built out of hard flinty sandstone. At the top of this giant wall, picture a great glacier with blue ice three hundred feet thick, crevassed and rent into a thousand yawning caverns, and crowding downwards, ever threatening to launch masses of ice large as great buildings into the valley below. Such avalanches are among the most sublime and thrilling spectacles that nature affords. The eye alone is incapable of appreciating the vast scale of them. The long period of silence at first and the thunder of the falling ice reverberated among the mountain-walls produce a better impression of the distance and magnitude.

I arrived at the lower end of the lake toward one o'clock. The lake was only disturbed in one long narrow strip toward the middle by a gentle breeze while all the rest was perfectly calm. This was one of those rare days of which each year only affords two or three, when the lake is calm at midday under a clear sky. The mirror surface of the water presented an inverted image of the mountains, the trees on the shore, and the blue sky. The true water surface and the sunken logs on the bottom of the lake joined with the reflected objects in forming a puzzling composite picture.

The brilliant sun had taken away the chill of morning

and coaxed forth a few forest birds, but there were no flowers or butterflies to recall real summer. It seemed as though this were the last expiring effort of autumn before the cold of winter should descend into the valley and with its finger on the lips of nature cover the landscape with a deep mantle of snow and bind the lake in a rigid layer of ice. Even at this warmest period of the day the sun's rays seemed inefficient to heat the atmosphere, while from the cold shadows of the forest came a warning that winter was lurking near at hand, soon to sweep down and rule uninterrupted for a period of nine long months.

CHAPTER III.

Surroundings of the Lake—Position of Mountains and Valleys—The Spruce and Balsam Firs—The Lyall's Larch—Alpine Flowers—The Trail among the Cliffs—The Beehive, a Monument of the Past—Lake Agnes, a Lake of Solitude—Summit of the Beehive—Lake Louise in the Distant Future.

AMONG the mountains on all sides of Lake Louise are many scenes of unusual beauty and grandeur. While the lake itself must be considered the focal point of this region, and is indeed wonderfully attractive by reason of its rare setting, the encircling mountains are so rough and high, the valleys separating them so deep and gloomy, yet withal so beautiful, that the scenery approaches perfection. The forces of nature have here wrought to their utmost and thrown together in apparently wild confusion some of the highest mountains in Canada and carved out gloomy gorge and rocky precipice till the eye becomes lost in the complexity of it all. Lakes and waterfalls reveal themselves among the rich dark forests of the valleys, and afford beautiful foregrounds to the distant snow mountains which seem to tower ever higher as one ascends.

A brief description of the topography in the vicinity of Lake Louise would be now in place. Southwestward from the lake is a range of very high and rugged mountains covered with snow and glaciers. This range is the crest of the continent of North America, in fact the great water-shed which divides the Atlantic and Pacific drainage. In this range are many peaks over 11,000 feet above sea level, an altitude which is near the greatest that the Rocky Mountains attain in this latitude. While farther south in Colorado there are scores of mountains 13,000 or 14,000 feet high, it must be remembered that no mountains in Canada between the International boundary and the railroad have yet been discovered that reach 12,000 feet. Nevertheless, these mountains of lesser altitude are far more impressive and apparently much higher because of their steep sides and extensive fields of perpetual snow.

This great range, forming the continental water-shed runs parallel to the general trend of the Rocky Mountains of Canada, or about northwest and southeast. Several spur ranges branch off at right angles from the central mass and run northeast five or six miles. Between these spur ranges are short valleys which all enter into the wide valley of the Bow. Lake Louise occupies one of these lesser valleys.

The several lateral valleys are all comparatively near Lake Louise and differ remarkably in the character of the scenery and vegetation. One is beautiful and richly covered with forests; another desolate and fearfully wild. The valley of Lake Louise contains in all three lakes, of which the smallest is but a mere pool, some seventy-five yards across.

Far up on the mountain side to the north of Lake Louise two little lakes were discovered many years ago. They are now to the visitor who spends but one day, almost the chief point of interest in this region. The trail thither leads into the dense forest from near the chalet and proceeds forthwith to indicate its nature by rising steadily and constantly. The tall coniferous trees cast a deep cool shade even on a warm day. So closely do the trees grow one to another that the climber is entirely shut out from the world of mountains and surrounded by a primeval forest as he follows the winding path. Among the forest giants there are two principal trees, the spruce and the balsam fir. Each is very tall and slender and at a distance the appearance of the two trees is closely similar. The spruce is the characteristic tree of the Rockies and is found everywhere. It reaches a height of 75 or 100 feet in a single tapering bole, closely beset with small short branches bent slightly downward, as though better to withstand the burden of snow in winter. In open places the lower branches spread out and touch the ground, but in forests they die and leave a free passage between the trees. The balsam tree is quite similar but may be discerned by its smoother bark which is raised from underneath by countless blisters each containing a drop of transparent balsam. Here and there are a few tall pines rivalling the spruces and firs in height but affording a strong contrast to them in their scattered branches and larger needles.

The ground is covered with underbrush tangled in a

dense luxuriance of vegetable life and partly concealing the ancient trunks of fallen trees long since covered with moss and now slowly decaying into a red vegetable mold.

At length, after half an hour of constant climbing, a certain indefinable change takes place in the forest. The air is cooler, the trees grow wider apart, and the view is extended through long vistas of forest trees. Presently a new species of tree, like our Eastern tamarack, makes its appearance. It is the Lyall's larch, a tree that endures the rigors of a subalpine climate better than the spruces and balsam firs, so that it soon becomes to the climber among these mountains an almost certain indication of proximity to the tree-line.

It is not far from the truth to say that the Lyall's larch is the most characteristic tree of the Canadian Rockies. It is not found in the Selkirk Range just west of the main range, and while it has indeed been found as far south as the International boundary, it has not been discovered in the Peace River valley to the north. Restricted in latitude, it grows on the main range of the Rockies only at a great altitude. Here on the borderland between the vegetable and mineral kingdoms it forms a narrow fringe at the tree-line and in autumn its needles turn bright yellow and mark a conspicuous band around all the cliffs and mountain slopes at about 7000 feet above sea level. Its soft needles, gathered in scattered fascicles, are set along the rough and tortuous branches, affording a scanty shade but permitting of charming glimpses of distant mountains, clouds, and sky among its gray branches and light-green foliage. It seems incapable of sending up a tall slender stem but branches out irregularly and presents an infinite variety of forms. Possibly for this reason the larch cannot contest with the slender spruces and firs of the valley, where it would be crowded out of light and sun among its taller rivals.

Presently the trail leads from out the forest and crosses an open slope where some years ago a great snow-slide swept down and stripped the trees from the mountain side. Here, 1200 feet above Lake Louise, the air feels sensibly cooler and indicates an Alpine climate. The mountains now reveal themselves in far grander proportions than from below, as they burst suddenly on the view. Nature has already made compensation for the destroyed forest by clothing this slope with a profusion of wild flowers, though much different in character from those at Lake Louise. Alpine plants and several varieties of heather, in varying shades of red or pink and even white, cover the ground with their elegant coloring. One form of heath resembles almost perfectly the true heather of Scotland, and by its abundance recalls the rolling hills and flowery highlands of that historic land. The retreating snow-banks of June and July are closely followed by the advancing column of mountain flowers which must needs blossom, bear fruit, and die in the short summer of two months duration. One may thus often find plants in full blossom within a yard of some retreating snow-drift.

On reaching the farther side of the bare track of the avalanche, the trail begins to lead along the face of craggy cliffs like some llama path of the Andes. The mossy ledges are in some places damp and glistening with trickling springs, where the climber may quench his thirst with the purest and coldest water. Wherever there is the slightest possible foothold the trees have established themselves, sometimes on the very verge of the precipice so that their spreading branches lean out over the airy abyss while their bare roots are flattened in the joints and fractures of the cliff or knit around the rocky projections like writhing serpents.

More than four hundred feet below is a small circular pond of clear water, blue and brilliant like a sapphire crystal. Its calm surface, rarely disturbed by mountain breezes, reflects the surrounding trees and rocks sharp and distinct as it nestles in peace at the very base of a great rock tower—the Beehive. Carved out from flinty sandstone, this tapering cone, if such a thing there be, with horizontal strata clearly marked resembles indeed a giant beehive. Round its base are green forests and its summit is adorned by larches, while between are the smooth precipices of its sides too steep for any tree or clinging plant. What suggestions may not this ancient pile afford! Antiquity is of man; but these cliffs partake more of the eternal—existing forever. Their nearly horizontal strata were formed in the Cambrian Age, which geologists tell us was fifty or sixty millions of years ago. Far back in those dim ages when the sea swarmed with only the lower forms of life, the fine sand was slowly and constantly settling to the bottom of the ocean and building up vast deposits which now are represented by the strata of this mountain. Solidified and made into flinty rock, after the lapse of ages these deposits were lifted above the ocean level by the irresistible crushing force of the contracting earth crust. Rain and frost and moving ice have sculptured out from this vast block monuments of varied form and aspect which we call mountains.

Just to one side of the Beehive a graceful waterfall dashes over a series of ledges and in many a leap and cascade finds its way into Mirror Lake. This stream flows out from Lake Agnes, whither the trail leads by a short steep descent through the forest. Lake Agnes is a wild mountain tarn imprisoned between gloomy cliffs, bare and cheerless. Destitute of trees and nearly unrelieved by any vegetation whatsoever, these mountain walls present a stern monotony of color. The lake, however, affords one view that is more pleasant. One should walk down the right shore a few hundred feet and look to the north. Here the shores formed of large angular blocks of stone are pleasantly contrasted with the fringe of trees in the distance.

The solitary visitor to the lake is soon oppressed with a terrible sensation of utter loneliness. Everything in the surroundings is gloomy and silent save for the sound

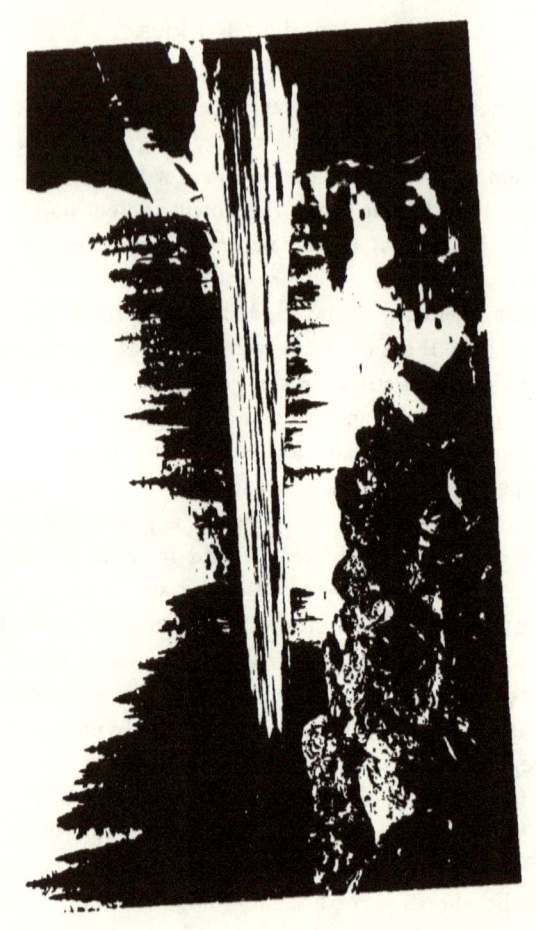

of a trickling rivulet which falls over some rocky ledges on the right of the lake. The faint pattering sound is echoed back by the opposite cliffs and seems to fill the air with a murmur so faint, and yet so distinct, that it suggests something supernatural. The occasional shrill whistle of a marmot breaks the silence in a startling and sudden manner. A visitor to this lake once cut short his stay most unexpectedly and hastened back to the chalet upon hearing one of these loud whistles which he thought was the signal of bandits or Indians who were about to attack him.

Lake Agnes is a narrow sheet of water said to be unfathomable, as indeed is the case with all lakes before they are sounded. It is about one third of a mile in length and occupies a typical rock basin, a kind of formation that has been the theme of heated discussion among geologists. The water is cold, of a green color, and so pellucid that the rough rocky bottom may be seen at great depths. The lake is most beautiful in early July before the snowbanks around its edge have disappeared. Then the double picture, made by the irregular patches of snow on the bare rocks and their reflected image in the water, gives most artistic effects.

From the lake shore one may ascend the Beehive in about a quarter of an hour. The pitch is very steep but the ascent is easy and exhilarating, for the outcropping ledges of sandstone seem to afford a natural staircase, though with irregular steps. Everywhere are bushes and smaller woody plants of various heaths, the tough strong branches of which, grasped in the hand, serve to assist the climber, while occasional trees with roots looped and knotted over the rocks still further facilitate the ascent.

Arrived on the flat summit, the climber is rewarded for his toil. One finds himself in a light grove of the characteristic Lyall's larch, while underneath the trees, various ericaceous plants suggest the Alpine climate of the place.

Though the climber may come here unattended by friends, he never feels the loneliness as at Lake Agnes. There the gloomy mountains and dark cliffs seem to surround one and threaten some unseen danger, but here the broader prospect of mountains and the brilliancy of the light afford most excellent company. I have visited this little upland park very many times, sometimes with friends, sometimes with the occasional visitors to Lake Louise, and often alone. The temptation to select a soft heathery seat under a fine larch tree and admire the scenery is irresistible. One may remain here for hours in silent contemplation, till at length the rumble of an avalanche from the cliffs of Mount Lefroy awakens one from reverie.

The altitude is about 7350 feet above sea level and in general this is far above the tree line, and it is only that this place is unusually favorable to tree growth that such a fine little grove of larches exists here. Nevertheless, the summer is very brief—only half as long as at Lake Louise, 1700 feet below. The retreating snow-banks of winter disappear toward the end of July and new snow often covers the ground by the middle of September. How could we expect it to be otherwise at this great height and in the latitude of Southern Labrador? On the hottest days, when down in the valley of the Bow the thermometer may reach eighty degrees or more, the sun is here never oppressively hot, but rather genially warm, while the air is crisp and cool. Should a storm pass over and drench the lower valleys with rain, the air would be full of hail or snow at this altitude. The view is too grand to describe, for while there is a more extensive prospect than at Lake Louise the mountains appear to rise far higher than they do at that level. The valleys are deep as the mountains high, and in fact this altitude is the level of maximum grandeur. The often extolled glories of high mountain scenery is much overstated by climbers. What they gain in extent they lose in intent. The widened horizon and countless array of distant peaks are enjoyed at the expense of a much decreased interest in the details of the scene. In my opinion one obtains in general the best view in the Canadian Rockies at the tree line or slightly below. Nevertheless every one to his own taste.

The most thrilling experience to be had on the summit of the Beehive is to stand at the verge of the precipice on the east and north sides. One should approach cautiously, preferably on hands and knees, even if dizziness is unknown to the climber, for from the very edge the cliff drops sheer more than 600 feet. A stone may be tossed from this place into the placid waters of Mirror Lake, where after a long flight of 720 feet, its journey's end is announced by a ring of ripples far below.

Lake Louise appears like a long milky-green sheet of water, with none of that purity which appears nearer at hand. The stream from the glacier has formed a fan-shaped delta, and its muddy current may be seen extending far out into the lake, polluting its crystal water and helping to fill its basin with sand and gravel till in the course of ages a flat meadow only will mark the place of an ancient lake.

There are even now many level meadows and swampy tracts in these mountains which mark the filled-up bed of some old lake. These places are called "muskegs," and though they are usually safe to traverse, occasionally the whole surface trembles like a bowl of jelly and quakes under the tread of men and horses. In such places let the traveller beware the treacherous nature of these sloughs, for on many an occasion horses have been suddenly engulfed by breaking through the surface, below which deep water or oozy mud offers no foothold to the struggling animal.

At the present rate of filling, however, the deep basin of Lake Louise will require a length of time to become obliterated that is measured by thousands of years rather than by centuries,—a conception that should relieve our anxiety in some measure.

CHAPTER IV.

Organizing a Party for the Mountains—Our Plans for the Summer—William Twin and Tom Chiniquy—Nature, Habits, and Dress of the Stoney Indians—An Excursion on the Glacier—The Surface Debris and its Origin—Snow Line—Ascent of the Couloir—A Terrible Accident—Getting Down—An Exhausting Return for Aid—Hasty Organization of a Rescue Party—Cold and Miserable Wait on the Glacier—Unpleasant Surmises—" I Think You Die"—A Fortunate Termination.

PREVIOUS to the summer of 1894 my experiences in the Canadian Rockies had made me acquainted with but little more of their general features and scenery than has been already described. This was sufficient, however, to prove that a most delightful summer could be spent among these mountains if a party of young men were organized with some definite object in view to hold the party together. Several of us accordingly assembled at one of our eastern colleges and discussed plans for the summer. Four men were persuaded to go on this excursion after the glories of the region had been duly set forth and the evidence corroborated so far as possible by the use of photographs. We were to meet at Lake Louise, where our headquarters were to be at the chalet, as near the first of July as possible.

Though the individual inclinations of the various members of our party might seem unlikely to harmonize together, we had nevertheless agreed on carrying out a certain plan. One of the party was an enthusiastic hunter, another eager for the glories of mountain ascents, one a geologist, another carried away by the charms of photography, while the fifth and last was ready to join in almost any undertaking or enterprise whatsoever.

However, our common purpose joined us all together to a certain degree. This was to explore and survey the region immediately around Lake Louise, to ascend several of the highest peaks, to get photographs of the best scenery, and in general to learn all we could about the environment of the lake.

Three of us arrived at the lake one fine morning early in July. The beauty of the scenery seemed to make a deep impression on my friends, and fortunately the clouds which at first concealed the mountain tops lifted soon after our arrival and produced very grand effects. At that time there were two Stoney Indians at the lake, who were engaged in cutting a trail to a lately discovered point of interest. One of these was named William Twin; his surname was probably derived from the fact that he had a twin brother, whose name was Joshua. A Stoney Indian who once acted as my guide was named Enoch; and upon being asked his surname he replied, "Wildman." These curious cases afford good examples of the origin of names. William was a fine-looking Indian. He came nearer to a realization of the ideal Indian features such as one sees on coins, or in allegorical figures, than almost any savage I have ever seen.

Tom Chiniquy was the other of the two Indians, and indeed the more important, as he is the eldest son of Chief Chiniquy, who in turn is under Bears' Paw, the head chief of all the Stoneys. An air of settled gravity, stern and almost bordering on an appearance of gloom, betokened his serious nature. I cannot but admire these Stoney Indians, free as they are from the vices of civilization, while still retaining many of the simple virtues of savage life.

As we saw the Indians every day we soon became acquainted with them, especially as William could talk quite intelligibly in English. The very first day of our arrival at the chalet the sharp eyes of the Indians, which seemed to be ever roving about in search of game, discovered a herd of goats on the mountain side. In vain did we try to see them, and at length, by means of a pair of powerful field glasses, they appeared as small white spots without definite forms, whereas to the Indians they were plainly visible. William was disgusted with us, and said, " White man no good eyes," in evident scorn.

With practice, our race can excel the Indians in every undertaking requiring skill, patience, or physical endurance, with the exception of two things in which they are infinitely our superiors. These are their ability to discover minute objects at great distances, and to read those faint and indefinite signs made by the passage of man or game through the forests or on the hard plains, where a white man would be completely baffled. A turned leaf, a bent blade of grass, a broken twig, or even the sheen on the grass, leads the swarthy savage unerringly and rapidly along, where the more intelligent but less observant white man can see absolutely nothing.

The Indian is said to be stolid and indifferent, while the hard labor which the squaws are compelled to undergo is always laid up against them as an evidence of their brutal character. But on the contrary this is their method of dividing labor, and a squaw whose husband is compelled to work about their camps is the subject of ridicule among the rest. The squaws do all the work which rationally centres around the camp-fire, just as our wives preside over our hearths and homes. The bucks provide the food, and should privation occur they will cheerfully share their last morsel with their wives and children, and, the more honor to them, they will do the same by a white man. The long and arduous labors of the chase, requiring the severest physical exertion, exhaust the strength, often while exposed to cold and rain for long periods of time. The bucks rightly consider their labor ended when they reach their camp, or "teepee" as they call them. Here the squaws preside and perform all the labor of cutting and cooking the meat, preserving and dressing the hides, and even gathering the firewood. They cut the teepee poles and set up their tents; and when not occupied with these more severe labors, they spend their time in making moccasins, weaving baskets, or fancy sewing and bead-work.

After all, the poor Indian is our brother, and not very unlike his civilized conqueror. One day William told me that the year before he had lost his squaw and four children by the smallpox, and that it had affected him so that

he could not sleep. In his own simple form of expression, it was most pathetic to hear him speak of this sad event, which evidently affected him deeply. " Me sleep no more now," he would say, "all time think me, squaw die, four papoose die, no sleep me. One little boy, me—love little boy, me—little boy die, no longer want to live, me."

We had the satisfaction of rendering a great service to William through his child, who was a bright and handsome little fellow. By some accident a splinter of wood had become lodged in the boy's eye. We were at length attracted by the peculiar actions of the little fellow, and upon inquiry found that he must have been enduring great pain, though without making a murmur of discontent. We took the matter in hand at once and sent him down to Banff, where, under skilful medical attendance, his eyesight, than which nothing is more dear to an Indian and which was totally gone in the affected eye and partially so in the other, was restored in a great measure. William was very grateful to us ever after, and on returning, some ten days later, delivered himself somewhat as follows : " Me say very much obliged. Three white men pretty good, I think."

The Stoneys are a remarkable tribe of Indians. Their headquarters is at a little place called Morley, about twenty miles east of the mountains on the plains. Here they are under the religious instruction of the Rev. Mr. McDougal. So far as the Indian is capable of receiving and following the precepts of Christianity, the Stoneys seem to have equalled or surpassed all other tribes. They are said to be great Bible readers, and they certainly show some familiarity with the Old Testament history, if we may judge by their custom of adopting Bible names. They have been taught a certain arbitrary code by which they can read and write in a simple manner, while many of them talk English if not fluently at least intelligibly.

Their manner of dress is a concession to their own native ideas and those of civilization, for while they invariably cling to moccasins and usually affect trousers cut from blankets with broad wings or flaps at the sides, their costume is not infrequently completed by some old discarded coat received by purchase or gift from the white man. These Indians rarely wear hat or cap, but allow their straight black hair to reach their shoulders and serve in place of any artificial protection. On either side of the face the hair is gathered into a braid so as to do away with the inconvenience of constantly pushing back their loose hair.

Dr. Dawson says that the Stoney Indians have very few names for the mountains and rivers, and that they have only inhabited this region for about forty years. The greater part of the Indian names for various features of the country are in reality Cree or their equivalents in Stoney. The Stoneys have recently incorporated the families of the Mountain Crees with their own. According to De Smet, both the Crees and the Stoneys migrated southward from the Athabasca region a few years before 1849, and it is probable that they entered this region about that time.

Palliser's Opinion.

I cannot conclude this digression on the Stoney Indians without quoting a few remarks from Captain Palliser's reports. Though written nearly forty years ago these facts are no less true than at that time.

"The members of the Stone tribe are hard workers, as their life is one requiring constant exertion and foresight. They travel in the mountains or in the forests along their eastern base, in parties of six or seven families. The young men are always off hunting in search of moose or other kinds of deer, or of the Rocky Mountain sheep. The old men busy themselves cutting out the travelling tracks through the woods, while the women pack and drive the few horses they use for carring their small supplies. They generally use skin tents stretched on a conical framework of poles, but their wigwams are much smaller than those of the Plain Indians. The women dress all the skins of the animals they kill into a soft leather, which, when smoked, is the material used throughout the whole country for making moccasins, most of the fine leather being obtained from the Stoneys. They are excellent hunters, and though as a rule small and feeble in body, are probably capable of more endurance than any other class of Indians. They make trustworthy guides, and, with a few exceptions, after some acquaintance with this tribe, you no more expect to be deceived, or told lies, as a matter of course, than you would in a community of white men."

So much for the Rocky Mountain Stoneys, or as they are sometimes called, the Assiniboines.

The completion of our party did not take place at the wished-for time, and for more than two weeks Mr. F. and Mr. H., and I were alone at the chalet. We commenced our surveying work by measuring a very accurate base line on the lake shore, and began training by making various moderate excursions on the mountain sides. On the third day, however, after our arrival the whole plan of our party came near having a most sudden and unwished-for termination, together with results which nearly proved fatal to one of the party. The accident and its attendant circumstances proved the most exciting episode in all our experiences, and as it most clearly illustrates the chief danger of climbing in the Canadian Rockies, I shall describe it in detail.

It happened in this manner. On the 13th of July, Mr. H., Mr. F., and I started to make an exploration of the glacier that is plainly visible from the chalet and which, some two miles distant, flows down from the snow fields and hanging glaciers of Mount Lefroy. This glacier is formed from two branches, which come in from the east, and uniting into one great stream, terminate about one mile above the head of the lake. The extreme length from the snout measured to the highest part of the glacier is about three miles, while the average width is less than one third of a mile.

The object of this excursion was in great part to gain a little knowledge of the use of rope and ice-axe, which we expected would be required in much of our subsequent work. There was no difficulty in the first part of this excursion, as a good trail leads round the lake and some half-mile beyond. There we forded the icy stream which comes from the glacier and pursued our way between the moraine and the mountain side for nearly a mile on the east side of the glacier. Our next move was to ascend the moraine, which was very steep and about a hundred feet high at this point. On arriving at the sharp crest of the moraine, we saw the great ice stream some fifty feet below, and so thoroughly covered with debris and boulders that the glacier was almost totally concealed. The passage down the moraine was very disagreeable, as the loose stones all scratched and polished by their former passage under the glacier were now rolling from under our feet and starting up great clouds of dust. Just below, at the border of the glacier, the water from the melting ice had converted the clay of the moraine into treacherous pools of bluish-gray mud, veritable sloughs of despond. At length, by the use of our ice-axes, we gained the firmer ice and with it the advantage of far more pleasant walking. We found the whole surface of the glacier literally covered with sharp stones and boulders of all sizes up to those which must have measured ten feet square by twenty feet long. They represented all sorts of formations, shales, limestones, and sandstones thrown down in wild disorder over the entire surface of the ice. All this material had been wrested from the mountain side far up the valley by frost and avalanche, and was now slowly moving toward the great terminal moraine. In one place a large area of nearly half an acre was strewed with giant blocks of a peculiar kind of rock different from all the rest, which apparently had come thundering down the mountain walls in one great rock-slide many years ago. Large flat slabs of shale were seen here and there supported on pillars of ice, showing how much the general surface of the glacier had wasted away under the influence of the sun's heat, while these pillars had been protected by the shade of the stone.

Advancing half a mile over the field of debris, we came gradually to where there were fewer stones, and at length reached almost pure ice. The question always arises where do all the boulders and pebbles that cover the lower parts of the glaciers come from ? In the upper parts of the glaciers or *névé* regions, where the snow remains perpetual and increases from year to year, the stones from the mountain sides are covered as they fall, and are at length buried deep and surrounded by ice as the snow becomes compressed and solidified. As the glacier advances down the valley and descends to lower altitudes, a level is at length reached where the snowfall of winter is exactly balanced by the melting of summer. This is the snow line, or rather this is the best place in which to locate such a variable level. Below this line the surface of the glacier melts away more than enough to make up for the winter fall of snow, and, as a result, the stones and debris buried in the ice gradually appear on the surface. In the Canadian Rockies near this latitude the snow line on northerly exposures, as judged by this method, is about 7000 feet above the sea, which is also just about the level called tree line.

In mountainous regions, where the climate is very dry, as in Colorado or in certain parts of the Andes, there is a great belt of several thousand feet between tree line and snow line where there is not sufficient moisture to allow of tree growth nor sufficient snowfall to form glaciers at all. In the Canadian Rockies the climate is moist enough to make these lines approach, and in the Selkirk Range and regions of extreme humidity the snow line is actually lower than the tree line.

We advanced slowly over the glacier and found much of interest on every side. The surface of the ice was at first comparatively smooth and channelled with small streams of pure water which flowed along with utmost rapidity but almost without ripples, as the smooth icy grooves seem adapted to every whim of the flowing water. At length the ice became more uneven and our passage was interrupted by crevasses, around which we had to thread our way by many a turn and detour. Most of them were, however, partly filled or bridged by snow and we found no particular difficulty in pursuing our way. About one o'clock we found ourselves at the base of Mount Lefroy, a little beyond the point where the two branches unite, and we held a consultation as to the plan of our farther advance. Mount Lefroy rises from the glacier in precipitous cliffs on every side, and we were even now under the shadow of its gloomy and threatening rock wall. There is no apparent method of scaling this mountain except by a long *couloir* or snow slope, which rises from the glacier and ascends nearly 1000 feet to a more gentle slope above the precipice. It was our intention to ascend this mountain, if possible, some time during the summer but the results of our first exploration for a favorable route rather inclined us to give up further attempts.

The result of our consultation was the decision to climb a short way up the *couloir* in order to see if it were possible to reach the gentle slope above. If this proved practicable, the ascent of the mountain was almost assured, as no great difficulties presented themselves above. Accordingly we commenced the ascent, all roped together in true Alpine fashion, and soon found the pitch so steep that our ice-axes rendered us much assistance in cutting steps. A number of great *schrunds* or horizontal crevasses often found on such slopes appeared to block our way, but as we approached we found a passage round every one. They were boat-shaped holes in the snow some forty or fifty feet deep and about the same width. The bottom of each appeared smooth and apparently of firm snow, so that they were not in reality very dangerous obstacles, as compared with the narrow and wellnigh unfathomable crevasses of an ordinary glacier.

Nevertheless, when we had reached a point several hundred feet above the schrunds and were on a steep slope of snow, my companions advocated taking to the rock ledges on the right of the snow, as they were altogether inexperienced in mountain climbing and felt somewhat nervous. We found the rock ledges practicable and quite easy except for a great number of loose stones which went rattling down as we advanced. We were in a

A Terrible Accident.

gloomy narrow gorge filled with snow and hemmed in on either side by cliffs which rose with almost vertical sides, here and there dripping with water from the snows above.

Whenever we paused for a momentary rest and the sliding, rattling stones ceased to fall, we were oppressed by the awful silence of this cheerless place of rocks and snow nearly 8000 feet above sea level.

It was while ascending these rock ledges that the accident occurred which came so near proving disastrous. There were a series of ledges from six to ten feet high alternating with narrow shelves where the slope was only moderately steep. The whole place was strewed with loose stones and boulders, some of which were so delicately poised that the slightest touch seemed sufficient to send them crashing down the cliff. At length a very dangerous looking stone of large size could be seen on the next shelf above us apparently just balanced in its precarious position, for the light could be seen underneath its base. H. followed me in safety around this great boulder which must have weighed more than half a ton. I was on the point of ascending the next ledge with the assistance of H. when we both heard a dull grating sound below, and turning, beheld the great boulder starting to roll over, and F. just below it and on the point of falling over the cliff. F. fell about ten feet to the next shelf where he was partially checked by the rope and prevented from falling farther. But to our horror the boulder, which had now gained considerable motion, followed after, and leaping over the ledge, for a short but awful moment it seemed to hang in mid-air, and then came down on F. with terrible force. It seemed impossible that there should be anything left of our poor friend. With a horrible crash and roar the great stone continued down the gorge, attended by a thousand flying fragments till the rocky cliffs echoed again.

After a momentary pause, unable to move and riveted to our places in horror, we hastily scrambled down to our companion who lay on the cliff insensible and bleeding. Our first efforts were to staunch his wounds with snow and then a hasty examination proved that though his hip appeared dislocated he had received probably no further serious injury. This escape appeared almost miraculous and it is probable that in the flying cloud of stones a smaller piece just happened to come under the great boulder and supported it partially at one end so that the full force of the blow was not felt. It was now half-past two in the afternoon and we were three hours' journey from the chalet with a man on our hands absolutely incapable of walking or even partially supporting his weight. It was evident that one of us must needs hasten back to the chalet for aid, but first it was necessary to get down the long snow-slope to the glacier.

Fortunately our rope was fully sixty feet long and after tying a loop under F.'s shoulders, I anchored myself securely with my ice-axe in the snow, and then lowered him rapidly but safely the length of the rope. H. then went down to F. and held him while I descended, and thus after twelve or fifteen repetitions of this proceeding we all landed in safety on the glacier. Having selected a place on the ice which was partially covered with a few small stones, we took off our coats and placed our wounded companion on this hard cold couch.

Carrying nothing but my ice-axe, I started for the chalet at once. The first part of the journey, while threading the crevasses, was slow and somewhat dangerous without the rope, but by running whenever practicable and pushing my energies to the utmost, I reached the chalet in one hour and ten minutes, or less than half the time required by us to come up in the morning. Unfortunately no one was at the chalet except Joe the cook. I however got him started immediately to cut two long, stout poles and a piece of canvas with which to make a litter. The two Indians were on the mountain side near Mirror Lake working on the trail and Mr. Astley, the manager of the chalet, was guiding some visitors to Lake Agnes. There was no other course open than to climb up after them, though I was quite exhausted by this time. I found William after twenty minutes of hard climbing and made him understand the situation at once. One must use a simple manner of speech as near like their own as possible, so I said to him—" William, three white men go up big snow mountain. Big stone came down, hurt one man. Tom, Mr. Astley, you—all go up snow mountain, bring white man back." William's face was a picture of horror, and he asked in anxiety—" Kill him?" I said no, but that he must hurry and get the other men. Dropping his axe, he ran off for the others in all haste, while I returned to the chalet and gathered sundry provisions and stimulants.

The rescuing party of four men was started in about thirty minutes, and taking the boat, rowed down the lake, till at last the small black speck on the water disappeared from our view as they neared the farther end.

A two-and-a-half mile ride on horseback brought me to the railroad station, where I sent a telegram to Banff for the Doctor. As there would be no train till the next morning I made arrangements for a hand-car to bring the Doctor up at once. A response soon came back that he was just about to start on his long ride of thirty-eight miles to Laggan.

Meanwhile poor F. and H. were having a miserable time of it on the glacier. The long hours rolled by one after another and no sign of aid or assistance was apparent. The days were still very long, but at length the declining sun sank behind the great ridge or mountain wall extending northward from Mount Lefroy. The glacier which imparts a chilly dampness even to the brilliancy of a mid-day sun now rapidly became cold in the lengthening shadows, and the surface waters began to freeze, while the deep blue pools of water shot out little needles of ice with surprising rapidity.

As they had seen me no more after I had disappeared behind a swelling mound of ice, they conjured up in their imaginations the possibility that I had fallen into some deep crevasse or had hurt myself on the treacherous moraine. At length, urged to desperate resolves, they formed a plan of leaving the ice by the nearest route, at whatever hazard to life and limb, rather than die of cold and exposure on the glacier. They had abundant opportunity for studying the grand phenomena of this Alpine region near at hand: the thundering avalanches from the cliffs behind them, and the cracking, groaning ice of the glacier as the great frozen stream moved slowly over its rocky uneven bed.

At length, to their great joy, they discerned by means of a field-glass which we had carried with us in the morning, the boat leaving the lake shore and slowly approaching. In half an hour the party reached the near end of the lake and were then lost to view for nearly two hours, till at length four little black dots appeared about a mile distant moving over the ice toward them.

The rescuing party did not reach them till seven o'clock, or more than four hours after the accident occurred. The return to the chalet was most exhausting to the men, especially to the Indians, whose moccasins afforded poor protection against the sharp stones and ice of the glacier.

Two section men came up from Laggan and met the party as they were returning, and afforded timely aid by their fresh strength. Poor F. was carried in a canvas litter hastily constructed and consequently not perfect in its results, as it only served to lift him a very little above the ground at the best and then where the ground was very smooth. William observed his haggard face and woe-begone appearance with concern and entertained the invalid at frequent intervals by such remarks as, "You think you die, me think so too." The rescuing party arrived at the chalet shortly after midnight, while the Doctor appeared an hour later. Each party had been travelling for the last five hours toward the chalet, and while one was accomplishing about three miles the other covered more than forty.

Fortunately there were no injuries discovered that would not heal in a few weeks, and through the influence of mountain air and perfect rest, recovery took place much more quickly than could be expected.

CHAPTER V.

Castle Crags—Early Morning on the Mountain Side—View from the Summit—Ascent of the Aiguille—An Avalanche of Rocks—A Glorious Glissade—St. Piran—Its Alpine Flowers and Butterflies—Expedition to an Unexplored Valley—A Thirsty Walk through the Forest—Discovery of a Mountain Torrent—A Lake in the Forest—A Mountain Amphitheatre—The Saddle—Impressive View of Mt. Temple—Summit of Great Mountain—An Ascent in Vain—A Sudden Storm in the High Mountains—Phenomenal Fall of Temperature—Grand Cloud Effects.

WHILE poor F. was recovering from his injuries, and before the two other men had arrived, H. and I carried on the work of surveying the lake, and made several interesting excursions on the adjacent mountain sides.

One fine cool morning, we went up the valley about half a mile beyond the end of the lake, and commenced an ascent of the sharp-crested ridge on the east side of the valley. This ridge forms a connection between the massive mountain on the left of the lake, known as Great Mountain, and a very high summit, crowned with a fine glacier, and named by some one Hazel Peak, which lies about two miles due south of Lake Louise. This connecting ridge we called Castle Crags, a name readily suggested by the irregular forms and outlines of the sharp needles and fingers, pointing heavenward, which adorned its highest crest, and seemed to represent the battlements and embrasures of some great castle. Several sharp columns of stone, with vertical sides, and narrow, graceful forms, rose up from this great parapet built by nature. Resembling feudal towers or donjons, they seemed by their great altitude to pierce the blue vault of heaven, and to dwarf by their proximity the snowy crest of Hazel Peak, which, in reality, is several thousand feet higher.

To ascend this ridge, and, if possible, gain the summit of one of these needles, from which we hoped to obtain a fine idea of the valley to the east, was the purpose of our excursion. The ascent proved easy almost from the start. On leaving the stream, which we crossed by means of some great trees, long since overcome by age or storm, and now serving as convenient bridges at frequent intervals, we commenced to ascend a long, even slope of limestone boulders, stable in position, and affording easy walking. The air was fresh and cool, for the morning sun was just rising over the crest of Castle Crags, while the rays of light seemed to skip from boulder to boulder, and, gently touching the higher points, left the others in shade. There were no bushes or tangled underbrush to impede our way, and so we had abundant opportunity to enjoy the beautiful flowers which cropped out in little patches among the yellow, gray, and cream-colored limestones. This was a mountain climb that proved thoroughly enjoyable, for all the conditions of atmosphere, of weather, and easy ascent were in our favor. There is a charm about the early morning hours among the high mountains. The bracing coolness of the air, as yet still and calm after the chill and quiet of night, the gradually rising sun and increasing light, the unusual freshness of the flowers and green vegetation, in their sparkling bath of dew, and the quiet calls of birds,—all seemed to herald the birth of a new day, far richer in promise than any heretofore. The afternoon, with its mellow light and declining sun, is like the calm, cool days of October, with its dusty foliage and sear leaves, brilliant in autumnal colors, but ever suggesting the approach of bleak winter, and pointing back to the glories of the past. The morning points forward with a different meaning, and hopefully announces the activity of another day, even as spring is the threshold and the promise of summer time.

As we advanced, and gradually increased our altitude, the plants and flowers changed in variety, character, and size, till at length we left all vegetation behind, and reached the bottom of a long, gentle slope of snow. The sun had not, as yet, touched the snow, and it was hard and granular in the frosty air. The first snow on a mountain climb is always pleasant to a mountaineer. To him, as, indeed, to any one, the summer snow-bank has no suggestion of winter, with its desolate landscapes and cold blasts, but rather of some delightful experiences in the mountains during vacation. These lingering relics of winter have little power to chill the air, which is often balmy and laden with the fragrance of flowers, in the immediate vicinity of large snow areas. The trickling rivulet, formed from the wasting snows of the mountain side, is often the only place where, for hours at a time, the thirsty climber may find a cold and delicious draught. Instead of destroying the flowers by their chilly influence, these banks of snow often send down a gentle and constant supply of water, which spreads out over grassy slopes below, and nourishes a little garden of Alpine flowers, where all else is dry and barren.

Arrived at the top of the long snow-slope, we found ourselves already nearly 3000 feet above the valley and not far below the crest of the ridge. A rough scramble now ensued over loose limestone blocks, where we found the sharp edges, and harsh surfaces of these stones, very hard on our shoes and hands. Upon reaching the crest, we beheld one of those fearfully grand and thrilling views which this portion of the Rocky Mountains often affords. The most conspicuous object in the whole view was the glacier, which descends from the very summit of Hazel Peak, at an altitude of more than 10,000 feet, and sweeps down in a nearly straight channel to the north, and in the course of but little more than a mile descends 4000 feet. A gloomy, narrow valley hems in its lower half, and on the side where we were, the precipice rose, in nearly perpendicular sides from the ice, far heavenward to where we stood. We launched a few large stones over the verge of the beetling precipice, and watched them descend in a few great leaps into the awful abyss, where they were broken into a thousand fragments on projecting ledges, or else, striking the glacier, continued their course till the eye could no longer follow them.

We were standing just at the base of one of the *aiguilles* which, from the valley, seem like sharp points of rock, but, now that we were near, proved to be about sixty feet high. This needle appeared to be precipitous and inaccessible on our first examination. But we discovered a narrow crevice or gully on the west side which apparently offered a safe method of ascent. I was soon near the top of the needle, but at the most difficult part, where only one small crack in the rock offered a good hand-hold, I was warned not to touch one side where the cliff seemed parted, and filled with loose material. Making a reconnaissance, I found the back of this same crag likewise separated a little from the solid rock, and the crevice partially disguised by loose stones and dirt, which had settled in and filled the hollow. This crag was about ten feet high and six or seven feet square, and though it seemed impossible to disturb so great a mass, I felt inclined to take the safer course and leave it entirely alone, so I scrambled up by a more difficult route.

Arrived on the top of the needle, I told H., who had remained below, to get under shelter while I should put this crag to the test. He accordingly found a projecting ledge of rock a little to one side, while I sat down and got a good brace and started to push with my feet against the top of the crag. A slight effort proved sufficient, and with a dull grating sound the great mass, which must have weighed about twenty-five tons, toppled slowly over on its base, and then fell with a fearful crash against the sides of the cliff, and commenced to roll down the mountain side like a veritable avalanche. Through the cloud of dust and flying stones I could faintly discern the features of my friend below, apparently much interested in what was going on. It was well that I had not trusted to this treacherous stone.

After I had pushed down most of the loose stones, H. came up and joined me on the summit of the *aiguille*. This needle had a blunt point indeed, for it proved to be a flat table about fifty feet long and ten feet wide. We were 8,700 feet above sea-level, and the wind was raw and chilly as it swept up from the valley and over this ridge. The sun had but little power to temper the air, and we soon started on our descent. In about five minutes we reached the top of the long snow-slope, where we enjoyed a glorious glissade and rapidly descended more than a thousand feet. The best manner of glissading is to stand straight up and slide on the feet, having one leg straight and the other slightly bent at the knee. Trailing the ice-axe behind as a precaution against too great speed, or to check the motion in case of a fall, the mountaineer can thus, in a few minutes, rapidly coast down long slopes which may have required hours of toil to ascend. Nothing in the experience of climbers is more exhilarating than a good glissade down a long snow-slope. The rush of air, the flying snow, and the necessity for constant attention to balance—all give a sensation of pleasure, combined with a spice of danger, without which latter almost all our sports and pastimes are apt to be tame. Do not many of our best sports, such as polo, horseback riding, foot-ball, yachting, and canoe sailing, gain some of their zest from a constant possibility of danger?

A few minutes of rapid descent down the limestone slope led us to a fine, small spring, which dashed in a score of small streamlets over some rocky ledges covered with moss and ferns. Here we sat down in the cool shade of the cliffs and ate our lunch. The air was now warm and still, because we were not far above the valley, and here, instead of seeking the warmth of the sun as we had done on the cold mountain summit, a brief three-quarters of an hour before, we now enjoyed the shade afforded by the rocks and forest near us. We reached the chalet in time for a second lunch, and, as in our mountain exercise we never found any meal superfluous, we were ready to present ourselves at the table at once.

On the 28th of July, W. arrived at the chalet, and, as A. had likewise appeared a few days previously, our party of five was now complete.

One of the first points which we decided to occupy in our surveying work was a high peak above Lake Agnes, called Saint Piran. This mountain is very easy to ascend and on several occasions we found ourselves on the summit for one purpose or another. The summit is far above tree line and, indeed, almost reaches the upper limit of any kind of plant growth. The rounded top is crowned with a great cairn, about ten feet high, which has been used as a surveying point some time in the past.

During the midsummer months this mountain summit is sparingly covered with bright flowers, all of an Alpine nature, dwarfed in size and with blossoms enormously out of proportion to the stems and leaves. There are several species of composites which rest their heads of yellow flowers almost on the ground, and a species of dwarf golden-rod about three inches high, with only two or three small heads on the summit of the stem; but the most conspicuous is a kind of moss pink, which is in reality a mountain variety of phlox. This plant grows in spreading mats upon the ground, with small, rigid, awl-shaped leaves gathered in tufts along the stem, while here and there are small bright blossoms of a pink color. Mr. Fletcher, who has spent some time in this region investigating the flowers and insects, once found a plant of the pink family on this mountain, which proved by its little joints to be more than one hundred years old.

One day I came up here alone, and on reaching the summit was surprised to find Mr. Bean, an entomologist, busily at work collecting butterflies. Mr. Bean has lived at Laggan for a number of years, and has made a most valuable collection of the insects, especially the butterflies and beetles, of all this region. Remarkably enough, it is on just such spots as this lofty mountain summit, 8600 feet above tide, that the rarest and most beautiful butterflies assemble in great numbers, especially on bright, sunny days. Here they are invited by the gaudy Alpine flowers, which have devoted all their plant energy to large blossoms and brilliant colors, so as to attract the various insects to them.

I was much interested in Mr. Bean's work, as he is the first pioneer in this field and has made many valuable discoveries. He showed me one butterfly of small size and quite dark coloring, almost black, which he said was a rare species, first discovered in polar regions by the Ross expedition, and never seen since till it was observed flitting about on this high peak, where arctic conditions prevail in midsummer. It is wonderful how the various species vary in color, form, and habit; some of the butterflies are very wild and shy, never allowing a near approach by the would-be collector; others are comparatively tame; and while some fly slowly and in a straight course, other species dart along most rapidly, constantly changing direction in sharp turns, and completely baffle all attempts at pursuit.

From the summit of this mountain we discovered a small lake in the valley to the west, and, as no one at the chalet had apparently ever visited the lake, or even known of its existence, we decided to make an excursion to this new region. Accordingly, a few days later, three of us started by the trail toward Lake Agnes, and after reaching a point about 600 feet above Lake Louise, we turned to the right and endeavored to make a traverse around the mountain till we should gain the entrance to the other valley. Our plan was not very good and the results were worse. For about two miles, the walking was along horizontal ledges of hard quartzite rock carpeted with grass and heaths, and occasionally made very difficult by the short dwarf spruces and larches which, with their tough elastic branches, impeded our progress very much. The day was unusually warm, and we were glad to reach at length a small patch of snow, where we quenched our thirst by sprinkling the snow on large flat stones, the heat of which melted enough to give us a small amount of muddy water. The roughness of the mountain and the nature of the cliffs now compelled us to descend near a thousand feet, and thus lose all the benefit of our first ascent. We were constantly advancing westward, hoping to come at length upon some stream that must descend from the valley of the little lake. Every valley in these mountains must have some stream or rivulet to drain away the water resulting from the melting snows of winter and the rains of summer, and we were certain that, if we continued far enough, we would finally discover such a stream. After our descent we proceeded through a fine forest, densely luxuriant, and in some places much blocked by prostrate trees and giant trunks, mossy and half decayed. The air seemed unusually dry, and our thirst, which had been only in part appeased by our draught at the snow-bank, now returned in greater severity than ever.

Suddenly we heard a distant sound of water, which, as we approached, grew still louder, till it burst into the full, loud roar of a beautiful mountain stream. The water was clear as crystal and icy cold, while nothing could exceed the graceful beauty of the many leaps and falls of the stream as it dashed over its rocky bed. Here we took lunch in a shady nook, seated on some rocky ledges at the edge of the water, surrounded on all sides by deep cool forests. How wild this little spot was! Though the railroad was less than two miles distant, probably no white man had ever seen this pleasant retreat where we were resting.

Had our excursion ended here, we should have been repaid for all the toil, heat, and thirst we had endured, by this single experience.

Nor was our pleasure over, for the stream, we knew, would prove a certain guide to the little lake, and, with the anticipation of soon reaching some enchanting bit of scenery when we should arrive at this sheet of water, we pursued our way along the series of falls and cascades by which our new-found stream leapt merrily down the mountain slope. Such is the charm of mountain excursions in these unexplored and little known wilds, for here, nature is ever ready to please and surprise the explorer by some little lake or waterfall or a rare bit of mountain scenery.

A COOL RETREAT IN THE FOREST.

Though we had stopped for luncheon at a place where the dashing water made several cascades and falls of exquisite beauty, we found a constant succession of similar spots, where I was often tempted to delay long enough to take photographs. As the stream thus descended rapidly, we found steep rock ledges, cut in giant steps and overgrown with thick moss till they were almost concealed from view, on either side of the mad torrent. These afforded us an easy method of ascent. The rocky formation of the stream bed revealed many different kinds of stone, conglomerates, shales, and quartzites, in clearly marked strata all gently dipping toward the south.

At length the woods opened up on either side, while, simultaneously, the slope decreased in pitch, and the stream ran over a bed of loose, rounded stones and boulders in the bottom of a shallow ravine. In a moment more we reached the lake, much more beautiful than our first view from St. Piran had led us to expect, but, also, much smaller in area. It was a mere pool, clear and deep, but intensely, blue in color and partially surrounded by a thin forest. Passing round the shores and up the valley, we found ourselves in some beautiful meadows, or rather moors, wherein streams of snow-water wandered in quiet, sinuous courses and gathered at length into the stream that feeds the lake. We came on a great number of ptarmigan—the high mountain species of grouse characteristic of this region,—which, with their young broods hardly able as yet to fly, were the most abundant signs of life that we found in this valley.

The Saddle.

A vast amphitheatre or cirque, with lofty, bare walls nearly free of snow, formed the termination of the valley. We were not compelled, however, to return over the same route as we had come, for we found an easy pass with a long gentle slope of snow on our left. This led us over the divide and, by a long steep descent, brought us to Lake Agnes, where we took advantage of the trail down the mountain side to the chalet.

Our attention was next turned toward the exploration of the mountains and valleys to the east of Lake Louise, which seemed to offer greater possibilities of grand scenery than those on the opposite side. Accordingly, we made several visits to a high upland park or alp, which was in reality a sort of depression between Great Mountain and a lesser peak to the east. This depression and the two mountains, one vastly higher than the other, resemble in outline, a saddle with pommel and crupper and suggested a name for the place which seems eminently appropriate. A trail now leads to the Saddle, and the place has proven so popular among tourists that it is frequently in use.

The Saddle is a typical alp, or elevated mountain meadow, where long, rich grass waves in the summer breezes, beautified by mountain flowers, anemones, sky-blue forget-me-nots, and scarlet castilleias. Scattered larch trees make a very park of this place, while the great swelling slopes rise in graceful curves toward the mountain peaks on either side.

But this is only the foreground to one of the most impressive views in the Rocky Mountains. To the eastward about three miles, on the farther side of a deep valley, stands the great mass of Mount Temple, the highest peak near the line of travel in the Canadian Rockies. This mountain stands alone, separated from the surrounding peaks of the continental watershed to which it does not belong. Its summit is 11,658 feet above the sea-level, while the valleys on either side are but little more than 6000 feet in altitude. As a result, the mountain rises over a mile above the surrounding valleys, a height which approaches the maximum reached in the Canadian Rockies. All sides of this mountain, except the south, are so precipitous that they offer not the slightest possible hope to the mountain climber, be he ever so skilful. The summit is crowned by a snow field or glacier of small size but of remarkable purity, since there are no higher cliffs to send down stones and debris to the glacier and destroy its beauty. On the west face, the glacier overhangs a precipice, and, by constantly crowding forward and breaking off, has formed a nearly vertical face of ice, which is in one place three hundred and twenty-five feet thick. I have seen passengers on the trains who were surprised to learn that the ice in this very place is anything more than a yard in depth, and who regarded with misplaced pity and contempt those who have any larger ideas on the subject.

Avalanches from this hanging wall of ice are rather rare, as the length of the wall is not great and the glacier probably moves very slowly. I have never had the good fortune to witness one, though the thunders of these ice falls are often heard by the railroad men who live at Lag-

gan, just six miles distant. They must indeed be magnificent spectacles, as the ice must needs fall more than 4000 feet to reach the base of the cliff. The compactness of this single mountain may be well shown, by saying that a line eight miles long would be amply sufficient to encircle its base, notwithstanding the fact that its summit reaches so great an altitude.

The strata are clearly marked and nearly horizontal, though with a slight upward dip on all sides, and especially toward the Bow valley, so that the general internal structure of the mountain is somewhat bowl-shaped, a formation very common in mountain architecture.

The surroundings of this great mountain are equally grand. Far below in the deep valley, the forest-trees appear like blades of grass, and in the midst of them a bright, foamy band of water winds in crooked course like a narrow thread of silver,—in reality, a broad, deep stream. A small lake, nestling among the dark forests at the very base of Mount Temple, is the most beautiful feature in the whole view. The distance renders its water a dark ultra-marine color, and sometimes, when the light is just at the proper angle, the ripples sparkle on the dark surface like thousands of little diamonds. On the right, an awful precipice of a near mountain looms up in gloomy grandeur, like the cliffs and bottomless abysses of the infernal regions pictured by Doré. This we called Mount Sheol.

One may ascend from the Saddle to the summit of Great Mountain in an hour. Mr. A. and I ascended this mountain in 1893, before there was any trail to assist us, and we had a very hard time in forcing our way through the tough underbrush, while below tree line.

In the course of a great many ascents of this peak I have had several interesting adventures. The view from the summit is so fine that I have made many attempts to obtain good photographs from this point. One day, after a period of nearly a week of smoky weather, the wind suddenly shifted, and, at about ten o'clock in the morning, the atmosphere became so perfectly clear that the smallest details of the distant mountains were distinct and sharp, as though seen through a crystal medium. This was my chance, and I proceeded at once to take advantage of it. I had a large 8 x 10 camera and three plate-holders, which all went into a leather case especially made for the purpose, and which was fitted out with straps, so that it rested between my shoulders and left both hands free for climbing. It weighed altogether twenty-four pounds. With lunch in my pocket, I set out from the chalet with all speed, so as to arrive on the summit before the wind should change and bring back the smoke.

I climbed as I had never climbed before, and though the day was hot I reached the Saddle in an hour, and, without a moment's pause, turned toward Great Mountain and commenced the long ascent of its rocky slope. In fifty-five minutes more I reached the summit and had ascended 3275 feet above Lake Louise. The air was still clear and offered every promise of successful photographs, even as I was unstrapping my camera and preparing to set it up for work. Suddenly, the wind shifted.

once more to the south and brought back great banks of smoke, which came rolling over the snowy crest of Mount Lefroy like fog from the sea. In five minutes all was lost. Mount Temple appeared like a great, shadowy ghost, in the bluish haze, and the sun shone with a pale coppery light. Such are the trials and tribulations of the climber in the Canadian Rockies.

One day at the end of August, H. and I ascended this mountain with our surveying instruments. The barometer had been steadily falling for several days, and already there were cumulus clouds driving up from the southwest in long furrows of lighter and darker vapors, which obscured the entire sky. A few drops of rain on the summit compelled me to work rapidly, but, as yet, there was no warning of what was in store.

After all the principal points were located we packed up our instruments and commenced a rapid descent to the Saddle. The slope is of scree and loose material, which permits a rapid descent at a full run, so that one may gain the Saddle in about fifteen minutes. Arriving there I paused to get a drink at a small stream under some great boulders, fed by a wasting snow-bank. H. had gone off toward the other side of the pass to get his rifle, which he had left on the way up.

Suddenly I heard a rushing sound, and, looking up, saw a cloud of dust on the mountain side and the trees swaying violently in a strong wind. A mass of curling vapor formed rapidly against the cliffs of Great Mountain, and a dull moaning sound, as of violent wind, seemed to fill the air. The sky rapidly darkened and black clouds formed overhead, while below them the thin wisps of scud rushed along and seemed white and pale by contrast.

I was no sooner up on my feet than the approaching blast was upon me, and with such unexpected force did it come that I was laid low at the first impulse. My hat went sailing off into space and was never seen more. The first shock over, I gained my feet again and started to find H. The air changed in temperature with phenomenal rapidity, and from being warm and muggy, in the space of about five minutes it grew exceedingly cold, and threatened snow and hail.

Though everything betokened an immediate storm and a probable drenching for us, I had time to notice a magnificent sight on Mount Temple. As yet there were no clouds on the summit, but, as I looked, my attention was called to a little fleck of vapor resting against the precipitous side of the mountain, half-way between summit and base. So suddenly had it appeared that I could not tell whether it had grown before my eyes or was there before. From this small spot the vapors grew and extended rapidly in both directions, till a long, flat cloud stretched out more than a mile, when I last saw it. The vapors seemed to form out of the very air where a moment before all had been perfectly clear.

Realizing that the sooner we started the better chance we should have of escape, we flew rather than ran down the trail, and were only overtaken by the storm as we approached the lake. The temperature had dropped so rapidly that a cold rain and damp snow were falling when we reached the lake. The boat had drifted from its moorings, and was caught on a sunken log some distance from the shore. I waded out on a sunken log, where I expected at any moment to slip from the slimy surface and take an involuntary bath in the lake. The boat was regained by the time H. had arrived a few minutes later and we reached the chalet thoroughly drenched.

Such sudden storms in the Canadian Rockies are rather rare, and are almost always indicated in advance by a falling barometer and lowering sky. I have never at any other time observed such a sudden fall in temperature, nor seen the clouds form instantaneously far down on the mountain side as they had done in this storm. The sudden rush of wind, the curling vapors, and flying scud afforded a magnificent spectacle on the Saddle, and one that was well worth the drenching we suffered in penalty.

CHAPTER VI.

Paradise Valley—The Mitre Glacier—Air Castles—Climbing to the Col—Dark Ice Caverns—Mountain Sickness—Grandeur of the Rock-Precipices on Mt. Lefroy—Summit of the Col at Last—A Glorious Vision of a New and Beautiful Valley—A Temple of Nature—Sudden Change of Weather—Temptation to Explore the New Valley—A Precipitate Descent—Sudden Transition from Arctic to Temperate Conditions—Delightful Surroundings—Weary Followers—Overtaken by Night—A Bivouac in the Forest—Fire in the Forest—Indian Sarcasm.

THE valley to the east of Lake Louise and parallel to it, we named Paradise Valley, on account of the elegant park-like effect of the whole place and the beauty of the vegetation. Our first entrance into this region and the discovery of the valley were partially accidental. In fact, we were making an expedition for the purpose of finding a practicable route up Hazel Peak, on the day when we were diverted from our original plan, and tempted to explore this hitherto unseen part of the mountains.

It came about somewhat in this manner. On the 30th of July, all but F., who was still lame from his accident, left the chalet carrying rope and ice-axes, with the intention of making explorations on the southern slopes of Hazel Peak. Our party, numbering four, left the chalet

The Mitre Glacier.

at a little after eight o'clock, with the intention of returning no later than five in the afternoon. Our equipment, beside our Alpine implements, consisted of a camera, a prismatic compass, and that which proved no less necessary, our lunches and a whiskey flask.

Taking the boat, we rowed to the other end of the lake, and then followed the same route as our party of three had taken on the disastrous expedition of July 13th, till we came to the junction of the two glacier streams. Here we turned toward the east, and followed the moraine of the wide glacier between Mount Lefroy and Hazel Peak.

The whole valley between was floored by a smooth, nearly level glacier, about a half mile wide and perhaps two miles long. Presently we were compelled to get on the ice as the moraine disappeared; so we put on the rope, and advanced with more caution. It was not long, however, before W., who was next to last in our line, broke through the bridge of a crevasse, despite our care, and sank to his shoulders. This member of our party was not versed in the art of snow-craft, and to him, every occurrence common to mountain experiences, and Alpine methods of procedure, were alike novel and terrible. In consequence, this accident fell more severely on him, but fortunately, he was extricated almost immediately by the use of the rope.

At the head of our valley was a remarkable, symmetrical mountain, resembling in general outline a bishop's mitre. From the glacier and snow-fields where we were walking, there rose on either side of the Mitre, steep snow-slopes, which terminated in lofty *cols* about 8500 feet above sea-level. That on the north side of the Mitre was exceedingly steep, and was rendered inaccessible by reason of a great crevasse, extending from the precipices on either side, clear across the snow-slope. This crevasse must have been nearly twenty-five yards in width and of great depth. At one side there still remained a thin bridge of snow, suspended, as it were, in mid-air over the awful chasm, as though to tempt climbers on to their instant destruction, or perhaps to a lingering death from cold and hunger.

The pass on our left appeared the more propitious and seemed to offer a possible route to the summit of the divide. We were anxious to get a view into the valley beyond, even though it were but for a few moments. The unknown regions on the other side of the pass had long been for me a favorite pleasure-ground of the imagination. Some fate had hitherto denied us any idea of the place beyond the vaguest suggestions. Several ascents, or partial ascents, of mountains on all sides of this unknown valley, had revealed the outlines of the surrounding mountains, but some intervening cliff or mountain range had always, with persistent and exasperating constancy, shut off all but the most unsatisfactory glimpses. Starting from these substantial foundations of reality, my imagination had built up a wide circular valley, surrounded on all sides by curious mountains of indefinite and ever changing outline and position. The picture always appeared in a gloomy, weird light, as though under a cloudy sky, or while the sun was near totally eclipsed. By some curious analogy, this faint illumination was similar to that which we always associate with the first creation of land and water; or far back in the geologic ages, when strange and hideous reptiles,—some flying in the murky air, some creeping amid the swampy growths of cycads, calamites, and gigantic tree ferns,—excite a strange thrill of pleasure and awe combined, as though the soul were dimly perceiving some new revelation of the universe, though but vaguely. In this weird, gloomy valley I wandered careless, in my imagination, many days and at many times, among forests infested by strange, wild animals, harmless like those of Eden, and by the shores of ever new, ever changing lakes and rivers.

So strong had this picture become that I felt the most intense anxiety to succeed in reaching the top of our pass, and gain at length a view of the reality, even at the risk of shattering these pleasant air castles, and annihilating, in a single instant, one of my best mental pleasure-grounds.

There were many dangers to be risked, however, and many obstacles to be overcome before this advantage might be gained. The steep slope was rendered formidable by reason of many great *schrunds*, or horizontal crevasses, caused by the ice of the glacier below, moving downward. In the intense cold of winter the moving ice becomes rigid and nearly stagnant, while the drifting snows accumulate, so as partly to fill these rents in the ice and bridge them over by cornices built out from one side or the other. When the increasing warmth of summer causes the ice to become plastic and to move more rapidly, these rents grow wider and the snow-bridges melt away and eventually fall into the crevasses so as to leave impassable chasms, dangerous to approach. Fortunately, it was not so late in the season that all the bridges were broken down, else we should have been completely defeated, for, on either side, the glacier was hemmed in by dangerous rock precipices. The south side of the glacier, moreover, was subject to frequent rock falls from the disintegrating cliffs of the Mitre. As we advanced over the extensive *névé*, the slope increased gradually but constantly, and soon became so steep that steps had to be cut, and great care was necessary not to slip. We crossed some of the schrunds by bridges of snow, where it was necessary to proceed with great caution, and, by sliding the feet along, apply the weight gently, lest the bridge should break through. We passed round others by walking along the lower edge or lip of the crevasse, which gave us a splendid but almost terrifying view of the gloomy caverns, extending down through the snow and ice to unknown depths. The dark-blue roofs of these crevasses were hung with dripping icicles, while from far below could be heard the sound of rushing, sub-glacial streams. Three hours of this slow, toilsome work were necessary to gain 1000 feet in altitude. We were now more than 8000 feet above the sea, and the atmosphere was raw and cold. Large damp flakes of snow and granular hail fell occasionally from a cloudy sky, silently and swiftly, through a quiet atmosphere. The whole horizon was bounded by high mountains, covered with glaciers and patches of snow, altogether barren and destitute of vegetation. Not a single tree or shrub, nor even a grassy slope at the far end of the great amphitheatre of mountain walls by which we were hemmed in, relieved the stern, cold monotony of the scene. So far as we might judge by our surroundings, we might have been exploring the lonely, desolate mountains of Spitzbergen, or some distant polar land, where frost and winter rule perpetual. Our progress up the slope of the glacier was very slow, as each step had to be cut out with the ice-axe. The pitch was so steep that a misstep might have resulted in our all sliding down and making further exploration of the *schrunds* below. The whole party was, in consequence, more or less affected by these cheerless circumstances, and became much depressed in spirit. As, however, the condition of the body is in great part responsible for all mental and moral ailments, so it was in our case. Had we been walking rapidly, so that the circulation of the blood had been vigorous and strong, both mind and body would have been in good condition, and the cold air, the snow, and bleak mountains would have been powerless to discourage. It is always at such times that mountain climbers begin to ask themselves whether the results are worth the efforts to attain them. Any one who has climbed at all, as we learn by reading the experiences of mountaineers, at many times has said to himself: "If I get home safely this time I shall never again venture from the comforts of civilization." The ancients, when in the thick of battle, or at the point of shipwreck, were accustomed to vow temples to the gods should they be kind enough to save them, but they usually forgot their oaths when safely home. Mountaineers in like manner forget their resolves, under the genial influence of rest and food, when they reach camp.

After many disappointments, we at last saw the true summit of our pass or *col* not far distant, and only a few hundred feet above us. A more gentle slope of snow, free of crevasses, led to it from our position.

Now that we were confident of success, we took this opportunity to rest by a ledge of rocks which appeared above the surrounding snow field. Here we regained confidence along with a momentary rest.

Nothing could surpass the awful grandeur of Mount Lefroy opposite us. Its great cliffs were of solid rock, perpendicular and sheer for about 2500 feet, and then sloping back, at an angle of near fifty degrees, to heights which were shut off from our view by the great hanging glacier. We could just catch a glimpse of its dark precipices, where the mountain wall continued into the unknown valley eastward, through a gorge or rent in the cliffs south of the Mitre. A magnificent avalanche fell from Mount Lefroy as we were resting from our severe exertion, and held our admiring attention for several moments. Another descended from the Mitre and consisted wholly of rocks, which made a sharp cannonade as they struck the glacier below, and showed us the danger to which we should have been exposed had we ascended on the farther side of the slope.

Discovery of a Beautiful Valley.

Having roped up once more, we proceeded rapidly toward the summit of the *col*, being urged on by a strong desire to see what wonders the view eastward might have in store. This is the most pleasurably exciting experience in mountaineering—the approach to the summit of a pass. The conquest of a new mountain is likewise very interesting, but usually the scene unfolds gradually during the last few minutes of an ascent. On reaching the summit of a pass, however, a curtain is removed, as it were, at once, and a new region is unfolded whereby the extent of the view is doubled as by magic.

We were, moreover, anxious to learn whether a descent into this valley would be possible, after we should arrive on the *col*. We were alternately tormented by the fear of finding impassable precipices of rock, or glaciers rent by deep crevasses, and cheered on by the hope of an easy slope of snow or scree, whereby a safe descent would be offered.

Proceeding cautiously, as we approached the very summit, to avoid the danger of an overhanging cornice of snow, we had no sooner arrived on the highest part than we beheld a valley of surpassing beauty, wide and beautiful, with alternating open meadows and rich forests. Here and there were to be seen streams and brooks spread out before our gaze, clearly as though on a map, and traceable to their sources, some from glaciers, others from springs or melting snow-drifts. In the open meadows, evidently luxuriantly clothed with grass and other small plants, though from our great height it was impossible to tell, the streams meandered about in sinuous channels, in some places forming a perfect network of watercourses. In other parts, the streams were temporarily concealed by heavy forests of dark coniferous trees, or more extensively, by light groves of larch.

This beautiful valley, resembling a park by reason of its varied and pleasing landscape, was closely invested on the south by a half circle of rugged, high mountains rising precipitously from a large glacier at their united bases. This wall of mountains, continuing almost uninterruptedly around, hemmed in the farther side of the valley and terminated, so far as we could see, in a mountain with twin summits of nearly equal height, about one mile apart. The limestone strata of this mountain were nearly perfectly horizontal, and had been sculptured by rain and frost into an endless variety of minarets, spires, and pinnacles. These, crowning the summits of ridges and slopes with ever changing angles, as though they represented alternating walls and roofs of some great cathedral, all contributed to give this mountain, with its elegant contours and outlines, the most artistically perfect assemblage of forms that nature can offer throughout the range of mountain architecture.

On the north side of this mountain, as though, here, nature had striven to outdo herself, there rose from the middle slopes a number of graceful spires or pinnacles, perhaps 200 or 300 feet in height, slender and tapering, which, having escaped the irresistible force of moving glaciers and destructive earthquakes, through the duration

A Temple of Nature.

of thousands of years, while the elements continued their slow but constant work of disintegration and dissolution, now presented these strange monuments of an ageless past. Compared with these needles, the obelisks and pyramids of Egypt, the palaces of Yucatan, or the temples of India are young, even in their antiquity. When those ancient peoples were building, nature had nearly completed her work here.

Beyond the nearer range of mountains could be seen, through two depressions, a more distant range, remarkably steep and rugged, while one particularly high peak was adorned with extensive snow-fields and large glaciers.

Almost simultaneously with our arrival on the summit of the pass, a great change took place in the weather. The wind veered about, and the clouds, which hitherto had formed a monotonous gray covering, now began to separate rapidly and dissolve away, allowing the blue sky to appear in many places. Long, light shafts of sunlight forced a passage through these rents, and, as the clouds moved along, trailed bright areas of illumination over the valley below, developing rich coloring and pleasing contrasts of light and shade over a landscape ideally perfect.

This beautiful scene, which has taken some time to describe, even superficially, burst on our view so suddenly, that for a moment the air was rent with our exclamations and shouts, while those who had lately been most depressed in spirit were now most vehement in their expressions of pleasure. We spent a half-hour on the pass and divided up our work, so that while one took photographs of the scenery, another noted down the angles of prominent points for surveying purposes, while the rest constructed a high cairn of stones, to commemorate our ascent of the pass.

Whatever may have been the mental processes by which the result was achieved, we found all unanimous in a decision to go down into the new valley and explore it, whatever might result. The cold, desolate valley on which we now turned our backs, but which was the route homewards, was less attractive than this unknown region of so many pleasant features, where even the weather seemed changed as we approached it.

It was now already two-thirty P. M. We were 8400 feet above sea-level and at an unknown distance from Lake Louise, should we attempt the new route. Another great mountain range might have to be passed before we could arrive at the chalet, for aught we knew. There were, however, fully six hours left of daylight, and we hoped to reach the chalet before nightfall.

A long snow-slope descended from where we were standing, far into the valley. This we prepared to descend by glissading, all roped together, on account of W., who was this day enjoying his first experience in mountain climbing. An unkind fate had selected him, earlier in the day, to break through the bridge of the crevasse and now doomed him to still further trouble, for we had no sooner got well under way in our descent, before his feet flew out from under him, and he started to slide at such a remark-

A Precipitate Descent.   95

able rate that the man behind was jerked violently by the rope, and, falling headlong, lost his ice-axe at the same time. With consternation depicted in every feature, our two friends came rolling and sliding down, with ever increasing speed, spinning round—now one leading, now the other, sometimes head first, sometimes feet first. The shock of the oncomers was too much for the rest of us to withstand, and even with our ice-axes well set in the soft snow, we all slid some distance in a bunch. At length our axes had the desired effect and the procession came to a standstill. It required some time to unwind the tangled ropes wherein we were enmeshed like flies in a spider's web, owing to the complicated figures we had executed in our descent. Meanwhile, a committee of one was appointed to go back and pick up the scattered hats, ice-axes, and such other wreckage as could be found.

The end of the descent was accomplished in a better manner, and in less than ten minutes we were 1500 feet below the pass. A short, steep scramble down some rocky ledges, where strong alder bushes gave good support for lowering ourselves, brought us in a few minutes to the valley bottom. At this level the air was warm and pleasant as we entered an open grove of larch and spruce trees. In the last quarter of an hour we had passed through all the gradations from an arctic climate, where the cold air, the great masses of perpetual snow, and bleak rocks, made a wintry picture, to the genial climate of the temperate zone, where were fresh and beautiful meadows enlivened by bright flowers, gaudy insects, and the smaller mountain animals. Humboldt has truly said : " In the physical as in the moral world, the contrast of effects, the comparison of what is powerful and menacing with what is soft and peaceful, is a never failing source of our pleasures and our emotions."

We followed a small, clear stream of an unusual nature. In some places it glided quietly and swiftly over a sloping floor of solid stone, polished and grooved in some past age by glaciers. A little farther on, the character of the mountain stream suffered a change, and the water now found its way in many sharp, angular turns and narrow courses by large square blocks of stone, for the most part covered by a thick carpet of moss, while between were deep pools and occasional miniature waterfalls.

Pursuing our way with rapid steps, for we were like adventurers in some fairy-land of nature, where every moment reveals new wonders, we came at length to an opening in the forest, where the stream dashed over some rocky ledges, that frost and age had rent asunder and thrown down in wild disorder, till the stream bed was fairly strewn with giant masses of sandstone. Some of these colossal fragments were apparently just balanced on sharp edges, and seemed ever ready to fall from their insecure positions. The variety and novelty of form presented by the falling water, as the streamlets divided here and united there, some over, some under, the stone bridges accidentally formed in this confusion of nature, aroused our greatest admiration.

As we advanced down the valley towards the north,

the outlines of the mountains changed, and we recognized at length the bare slopes of the southern side of Mount Temple, which at first seemed to us a strange mountain. Meanwhile, we had approached very near to the base of the beautiful mountain with the double peak and the many pinnacles, and found that proximity did not render it less attractive.

The stream which we followed had been joined by many other rivulets and springs till it grew to be wide and deep. At length a muddy torrent, direct from the glacier at the head of the valley, added new volume and polluted the crystal snow-waters of the stream which we had followed from its very source.

For many hours we followed the banks of the small river formed by these two branches, and found it an almost continuous succession of rapids, constantly descending, and with a channel swinging to right and left, every few hundred yards, in a winding course.

H. and I led the way, and frequently lost sight of the others who were beginning to tire and preferred a slower pace. We waited on several occasions for them to come up with us, though it seemed as if we should no more than reach the chalet before nightfall, even by putting forth our best efforts.

About 6.30 o'clock we came to a swampy tract, where the trees grew sparingly, and gave the appearance of a meadow to an expanse of nearly level ground, covered with fine grass and sedges. Here, after a long wait for our friends, who had not been seen for some time, we decided to write a note on a piece of paper and attach it to a pole in a conspicuous place where they could not fail to see it. The mosquitoes were so numerous that it was almost impossible to remain quiet long enough to write a few words explaining our plans. On the top of the stick we placed a small splinter of wood in a slit, and made it point in the exact direction we intended to take.

Having accomplished these duties in the best manner possible, we set out for the chalet with all speed, as we did not relish the idea of making a bivouac in the woods and spending a cheerless night after our long fast. It was evident that we were now at the outlet of the valley, and that, unless we should encounter very rough country with much fallen timber, our chances were good for reaching the chalet before darkness rendered travelling impossible. It was likewise important to reach the lake on account of those at the chalet, who might think that the whole party had met with some accident on the mountain, unless some of us turned up that night.

We accordingly walked as fast as our waning strength permitted, and after surmounting a ridge about 800 feet high, which formed part of the lower slopes of Saddle Mountain, we found no great difficulty in forcing a passage through the forest for several miles, when we came upon the trail to the Saddle. We reached the lake at 8.15 P.M., and after shouting in vain for some one to send over a boat, we forded the stream and entered the chalet, where a sumptuous repast was ordered forthwith, and to which we did ample justice after our walk of twelve hours duration.

A Bivouac in the Forest.

Our less fortunate friends did not appear till the next morning. They discovered our note, but decided not to take our route, as they thought it safer to follow the stream till it joined the Bow River. They had not proceeded far, however, beyond the place where we had left the note, before they became entangled in a large area of fallen timber and prostrate trees, where they were overtaken by night and compelled to give up all hope of reaching Lake Louise till the next day. In the dark forest they made a small fire, and were at first tormented by mosquitoes and, later, by the chill of advancing night, so that sleep was impossible. The extreme weariness of exhausted nature, crowned by hunger and sleeplessness amid clouds of voracious mosquitoes, was only offset by the contents of a flask, with which they endeavored to revive their drooping spirits, and cherish the feeble spark of life till dawn.

Fortunately, the nights in this latitude are short, and at four o'clock they continued their way to the Bow River, which they followed till they reached Laggan.

About six days later, a little column of smoke was observed rising from the forests towards the east, and from Laggan we learned that the woods were on fire, and that about forty acres of land were already in a blaze. A large gang of section men were despatched at once with water buckets and axes to fight the fire. The fire did not prove so extensive, however, as at first reported, and in about two days all the men were recalled.

William said to one of us : " Me think two white man

light him fire"; to which our friends replied that it was impossible, as the fire had broken out nearly a week after they had been there.

William replied, with the only trace of sarcasm I have ever known him to use: "White man no light fire, oh no, me think sun light him."

CHAPTER VII.

The Wild Character of Paradise Valley—Difficulties with Pack Horses—A Remarkable Accident—Our Camp and Surroundings—Animal Friends—Midsummer Flowers—Desolation Valley—Ascent of Hazel Peak—An Alpine Lake in a Basin of Ice—First Attempt to Scale Mt. Temple—Our Camp by a Small Lake—A Wild and Stormy Night—An Impassable Barrier—A Scene of Utter Desolation—All Nature Sleeps—Difficulties of Ascent—The Highest Point yet Reached in Canada—Paradise Valley in Winter—Farewell to Lake Louise.

OUR delightful experience in Paradise Valley convinced us that a camp should be established in it near the southern base of Mount Temple, which we hoped to ascend. From this camp we intended to make branch excursions in all directions and learn something of the mountains toward the east and south. All this region, though so near the railroad, had apparently never been explored by the surveyors, and the early expeditions had of course never approached this region nearer than the Vermilion Pass on the east and the Kicking Horse Pass on the west. In all our expeditions through these lonely but grand mountain valleys, we never discovered any mark of axe or knife on the trees, any charred pieces of wood to indicate a camper's fire, nor any cairn or pile of stones to prove some climber's conquest.

In fact, the impenetrable barrier of mountains at every valley end dissolved the surveyor's hopes, even from a distance, of finding any practicable pass through the maze of lofty mountains and intervening valleys blocked with glaciers and vast heaps of moraine. The lone prospector would not be tempted by any sign of gold in the streams to explore these valleys, though the Indian hunter may have occasionally visited these regions in search of bears or the mountain goat.

We first blazed a trail from the chalet to the entrance of Paradise Valley. The route followed was merely the best and most open pathway that we could find through the forests, and though not more than three miles in length, it required as many hours to reach the valley entrance. Pack horses we obtained at the chalet, but no man could be found who would consent to act as our cook or assistant in managing the horses.

Our camp was at length established by the side of a small rivulet on the lower slopes of Mount Temple, where we found the altitude to be 6900 feet above sea-level. Our experiences with pack animals were of a most exciting nature and sometimes severely trying to our temper and patience. The horses were not accustomed to this service and performed all sorts of antics, smashing the packs among the trees, jumping high in air to clear a small stream six inches wide, or plunging regardless into rivers where, for a moment, the horse and packs would be submerged in the water. There was one place about two miles within the valley entrance that might well try the patience of Job himself. On one side of the stream, was an impassable area covered with tree trunks crisscrossed and piled two or three deep by some snow-slide of former years. On the other side of the stream, which we were compelled to take, was a dense forest. Below was a tangled growth of bush, and many fallen trees, all resting on a foundation of large loose stones covered six inches deep with green moss. Between these stones were deep holes and occasional underground streams, the water of which could be faintly heard below and which had probably washed away the soil and left these angular stones unprotected. To lead a horse through this place required the greatest skill, patience, and even daring. Without some one to lead the animal with a rope, the poor beast would stand motionless, but to pick one's way over the rough ground while leading the horse invariably ended in disaster. The very first hole was enough to frighten the horse, so that, instead of proceeding more slowly, the animal usually made a mad rush forward regardless of the leader, who invariably fled and sought the protection of a tree, while the horse soon fell prostrate among the maze of obstacles. In these frantic rushes many of us were several times trampled on by the horse, and the packs were smashed against the branches and trunks of trees, or torn off altogether. This was an exceedingly dangerous bit of ground, and it was remarkable that on so many occasions we were able to lead our horses through it without a broken leg.

One of our most remarkable adventures with a horse

may indeed test the credence of the reader, but five men can vouch for its actual occurrence. We were passing along through the forest in our usual manner, which was the outgrowth of much experience. First of all, one man preceded and did nothing else but find the blaze marks and keep on the ill-defined trail as well as possible. About twenty-five yards behind came another man whose duty it was to find the pathfinder, and if possible, improve on his trail. Then came one of our party who led the horse with a long head rope, while behind the horse were two men whose duty it was to pick up whatever articles fell out of the packs from time to time, and fasten them on again.

As we were proceeding in this manner, we came to a slanting tree which leaned over the trail at an angle of about thirty degrees. It was just small enough to be limber, and just large enough to be strong. Moreover, it was too low for the horse to go under, and a little too high for him to jump over. One might travel a lifetime and never meet with just such another tree as this. In less than ten seconds this tree had brought the horse and two of our party to the ground and wrought consternation in our ranks.

As the horse approached the slanting tree, F., who was leading, saw the animal rear high in the air to prepare for a jump. He thought it best to get out of the way, but in his haste stumbled and fell headlong into a bush. Meanwhile the horse, a stupid old beast, prepared for the effort of his life, and with a tremendous spring jumped high in air, but unfortunately his fore-feet caught on the small tree, which swung forward a little and then returning like a powerful spring, turned the animal over in mid-air. The horse landed on his back some five yards farther on, and, with his four legs straight up in the air, remained motionless as death. But this was not all, for the tree swung back violently and struck H. on the nose, fortunately at the end of the swing, but with sufficient force to knock him down.

When our two friends recovered, we turned our attention to the horse, which still remained motionless on his back. "He is dead," said F., but, on rolling him over, the poor animal got up and seemed none the worse for his experience, except for a more than usual stupidity.

We camped about ten days in Paradise Valley in a beautiful spot near the end. Here, on all sides except towards the north, the place is hemmed in by lofty mountains. We saw the valley in all sorts of weather, in clear sunshiny days, and when the clouds hung low and shut out the mountains from view. On one or two occasions the ground was white with snow for a short time, though our visit was during the first part of August.

Many kinds of animals frequented the valley, and some of the smaller creatures lived in the rocks on all sides of our camp and became quite friendly. One of the most interesting little animals of the Canadian Rockies is the little pica, or tailless hare. This small animal abounded in the vicinity of our camp and is in fact always found at about 7000 feet altitude. It is a hare about the size of a rat, which, with its round ears, it more resembles. These little fellows have a dismal squeak, and they are very impertinent in their manner of sitting up among the rocks at the entrance to their holes, and gazing at their human visitors, ever ready to pop out of sight at a sign of danger. Chipmunks were likewise abundant and visited our camp to pick up scattered crumbs from our table.

There is a species of rat with a bushy tail that lives in the forests and rocky places of these mountains and is the most arrant thief among all the rodents. Nothing is too large for them to try and carry off, and they will make away with the camper's compass, aneroid, or watch, and hide them in some inaccessible hole, apparently with the desire to set up a collection of curios.

The siffleur, or marmot, is the largest among these rodents, and reaches the length of twenty-five or thirty inches. These animals usually frequent high altitudes at, or above the tree line, where they build large nests among the rocks and lay up a store of provisions for winter time. They are very fat in the fall, but it is not known whether they hibernate or not. Their note is a very loud shrill whistle, which they make at a distance, but they never allow one to approach very near, like the impudent picas.

We saw very few of the mountain goats, though we often came upon their fresh tracks in the mud near streams or in the snow far up on the mountain sides. On several occasions we could hear the patter and rattle of stones sent down by the movements of some herd, though our eyes failed to detect them.

Where the forests grew thick in the valley, the herbs and flowering plants were always less numerous, but in the meadows the ground was colored by mountain flowers of beautiful shades and pretty forms. The tasselled heads of the large anemones, long since gone to seed, were conspicuous everywhere, and they are always a beautiful object among the meadow grass as the summer breezes make gentle waves over these seas of verdure. Along the bare rocky margins of the streams, where all else has been forced to retire by occasional floods, two species of plants make a most brilliant coloring and dazzle the eye with discordant shades. They are the castilleias, or painter's brush, with bright scarlet and green leaves clustered at the top of a leafy stem, and the epilobiums, with reddish-purple blossoms; these two plants were often so close together with their inharmonious color tones as to perplex the observer in regard to nature's meaning. When nature does such things we grow to like her apparent mistakes, just as we love the bitter-sweet chords of Schumann, or Grieg's harsh harmonies.

We made several excursions into the next valley to the eastward, and beyond that, over the water-shed into British Columbia. The valley to the east offered the greatest contrast to Paradise Valley. It was somewhat wider, the altitude was in general higher, so that a great part was above the tree line, while the awful wildness and confusion created by vast heaps of moraine and a large glacier at the foot of a range of saw-edged mountains made this place seem like a vale of desolation and death.

At the close of our camping experiences, we effected the conquest of two mountains, Hazel Peak and Mount Temple, on two successive days. We first tried Hazel Peak, and by following the route which had been previously selected, we found the ascent remarkably easy. On the summit, the climber is 10,370 feet above sea-level,—higher than the more celebrated Mount Stephen, often claimed to be the highest along the railroad,—and surrounded by more high peaks than can be found at any other known part of the Canadian Rockies, south of Alaska. In fact there are seven or eight peaks within a radius of six miles that are over 11,000 feet high.

The view is, at the same time, grand and inspiring, and has certain attractions that high mountain views rarely present. The rock precipice and snow-crowned crest of Mount Lefroy are separated from the summit of Hazel Peak by one of the grandest and deepest canyons of the Canadian Rockies, so that the distance from summit to summit is only one mile and a half. The ascent of Hazel Peak is certainly well worth the labor of the climb, as the round trip may be easily accomplished from Paradise Valley in five hours, though the ascent is nearly 4000 feet.

On the north side, from the very summit, a fine glacier sweeps down in steep pitch far into the valley below and with its pure white snow and yawning blue crevasses of unfathomable depth, forms one of the most attractive features of this mountain. The most remarkable and beautiful object that we discovered, however, was a small lake or pool of water only a few yards below the summit

First Attempt on Mt. Temple.

of the mountain. Encircled on all sides by the pure snows of these lofty altitudes, and embedded, as it were, in a blue crystal basin of glacier ice, the water of this little lake was colored deep as indigo, while over the surface a film of ice, formed during the previous night, had not yet melted away.

We returned to camp much elated with our success but doubtful of the morrow, as no easy route had yet been discovered up the forbidding slopes of Mount Temple. The year before, Mr. A. and I had been hopelessly defeated even when we had counted most on success. Moreover, the mere fact that, though this mountain was the highest yet discovered anywhere near the railroad, it had never been ascended by any surveyor or climber, made success appear less probable, though it urged us on to a keener ambition.

The attempt by A. and myself to ascend this mountain in 1893 was probably the first ever made. During the first week of August, we started from Laggan, having with us a Stoney Indian, named Enoch Wildman, and one horse to carry our tent and provisions. The day was unusually hot, and, as we forced our monotonous and tiresome passage through the scanty forests of pine near the Bow River, we suffered very much from heat and thirst. In these mountain excursions, it is the best policy to wear very heavy clothes, even at the disadvantage of being uncomfortable during the day, for the nights are invariably cold, even at low altitudes. We did not camp until nightfall, when we found ourselves on the northern slope of the mountain, 7000 feet above sea-level, by the side of a small lake. The little lake occupied a depression among giant boulders and the debris of the mountain. At one end, a large bank of snow extended into and below the water, which was apparently rising, as there were fragments of frozen snow floating about in the lake. The banks sloped steeply into the water on all sides, and there was not a single level spot for our camp, so that it was necessary to build a wall of stones, near the water's edge, for our feet, and to prevent ourselves from sliding into the lake during the night.

The weather was wild and stormy, and the long night seemed to drag out its weary length to an interminable extent of time, attended as it was by showers of rain and hail and furious gusts of wind, which threatened to bring our flapping tent to the ground at any moment.

Our camp-fire, which had been built on a scale appropriate to some larger race of men, was a huge pile of logs, each fully ten feet long, and twelve or eighteen inches through, but the wind blew so strong that the mass roared like a vast forge during the early hours, and then died away into an inert mass of cinders toward the chill of morning.

The light of day revealed our wild surroundings. We were under the northern precipice of Mount Temple, and so close that we could see only the lower part of this inaccessible wall. A beautiful fall dashed down in a series of cascades through a distance of about 1000 feet, and fed our little lake. Sometimes the strong wind, blow-ing against the cliff, or sweeping upward, would make the water pause and momentarily hang in mid-air, suspended, as it were, on an invisible airy cushion, till gathering greater volume, it would burst through the barrier and fall in a curtain of sparkling drops.

Poor Enoch had suffered terribly from cold during the night, and begged our permission to return to Laggan, promising to come back the next day—"sun so high," pointing to its place in the early afternoon. He said in his broken English: "No grass for pony here, too cold me; no like it me." So we took pity on him and sent him back to more comfortable quarters while we rested in comparative quiet, it being Sunday.

Early Monday morning we had our breakfast and were on foot at four o'clock. The gloom of early dawn, the chill of morning, and the cloudy sky had no cheering effect on our anticipations. Our plan was to traverse the mountain side till we should come to the southeast shoulder, where we had once observed an outline of apparently easy slope.

By eleven o'clock we had reached an altitude of nearly 10,000 feet without meeting with any very great difficulty, but here we came suddenly to a vertical wall of rock about 400 feet high and actually leaning over in many places, a barrier that completely defeated us, as the wall extended beyond·our view and offered no prospect of giving out. At the base of this cliff was a steep, narrow slope of loose, broken limestone, and then another precipice below. Along this dangerous pathway we continued for some distance, keeping close to the base of the cliff. The loose stones, set in motion by our feet, slid down and rolled over the precipice, where we could hear them grinding to powder on the cliffs below.

Never in my life have I been so much impressed with the stern and desolate side of nature. The air was bitter cold and had the frosty ozone odor of winter. A strong wind rushed constantly by us, and, as it swept up the gorges of the precipice above, and over the countless projections of the cliffs, made a noise like the hoarse murmur of wind in a ship's rigging, or the blast of some great furnace. To the south and east, range beyond range of bare, saw-edged mountains raised their cold, sharp summits up to a cloudy sky, where the strong wind drove threatening clouds in long trains of dark and lighter vapors. The intervening valleys, destitute of vegetation or any green thing, were filled with glaciers and vast heaps of moraine, and the slides of debris from the adjacent mountain side. All was desolate, gloomy, cold, and monotonous in color. Three thousand feet below, a small lake was still bound fast in the iron jaws of winter, surrounded as it was by the walls of mountains which shut out the light and warmth of the summer sun. Inert, inanimate nature here held perpetual rule in an everlasting winter, where summer, with its flowers and birds and pleasant fertility, is unknown, and man rarely ventures.

Overcome with the terrors of this lonely place and the hopelessness of further attempt to reach the summit, where a snow-storm was now raging, we turned back. As we reached our camp we found Enoch just approaching, according to his promise, and though the afternoon was well advanced, we packed up and moved with all speed toward Laggan. We reached Lake Louise at 10.30 P.M., after almost nineteen hours of constant walking.

Now, however, at our camp in Paradise Valley, the conditions were somewhat different. We were at the very base of the mountain, and had learned much more about it, in the year that had elapsed since our first attempt.

The mountaineer has many discomforts mingled with the keen enjoyment of his rare experiences. None is more trying than the early hour at which he is compelled to rise from his couch of balsam boughs and set forth on his morning toil. At the chill hour before dawn, when all nature stagnates and animate creation is plunged in deepest sleep, the mountain climber must needs arouse himself from heavy slumber and, unwilling, compel his sluggish body into action.

This is the deadest hour of the twenty-four—the time just before dawn. The breezes of early night have died away into a cold and frosty calm; the thermometer sinks to its lowest point, and even the barometer, as though in sympathy, reaches one of its diurnal minima at this untimely hour. And if inanimate nature is thus greatly affected, much more are the creations of the vegetable and animal kingdoms. The plants are suffering from the cold and frost; the animals of daytime have not as yet aroused themselves from sleep, while the nocturnal prowlers have already ceased their quest of prey and returned to their dens. Even man is affected, for at this dead hour the ebb and pulse of life beat slow and feeble, and the lingering spark of life in those wasted by disease comes at this time most near going out.

At such an unseasonable hour, or more accurately at four A.M., were we up, on the 17th of August preparing for our ascent of Mount Temple. There was no trace of dawn, and the waning moon, now in her last quarter, was riding low in the southern sky, just above the sharp triangular peak at the end of our valley.

At nine o'clock in the morning, we had gained the summit of the pass between Mount Temple and Pinnacle Mountain, where we were 9000 feet above sea-level. The ascent so far had not been of an encouraging nature, as we had encountered a long, loose slide where everything moved threateningly at each step. I have never seen a more unstable slope. The stones and boulders would slide, and begin to move at a distance of ten and fifteen feet above the place where we stood, and on every side also. F., who was one of the party, was terror-stricken, for he now had a horror of moving stones of any description.

The view from this pass was very extraordinary. To the east stood the rugged, saw-edged mountains of the Desolation Range, looming up in solemn grandeur through an atmosphere bluish and hazy with the smoke of forest fires. The air was perfectly calm and had the bracing coolness of early morning and high altitude, which the rising sun tempered most gently. The weather conditions for accomplishing our ascent were perfect, but there was little prospect of a fine view by reason of the smoke.

The outlook from the pass was indeed discouraging. Cliffs and ledges with broken stones and loose debris seemed to oppose all safe passage. Fortunately, as we progressed the difficulties vanished, and not till we reached an altitude of about 10,000 feet did we encounter any real obstacles. We found a passage through the great rock wall which had defeated us last year, by the aid of a little gully, which, however, entailed some rather difficult climbing. This arduous work continued throughout the next 1000 feet, when, at an altitude of 11,000 feet, we came to the great slope between the southwest and west *arêtes* and found an easy passage to the summit.

SUMMIT OF MOUNT TEMPLE.

Many a hearty cheer rent the thin air as our little party of three reached the summit, for we were standing where no man had ever stood before, and, if I mistake not, at the highest altitude yet reached in North America north of the United States boundary. The summit was formed of hard bluish limestones, broken and piled up in blocks, as on all high mountain tops. The cliffs toward the east were stupendous and led the eye down to the valley more than a mile below. The air was almost calm and just above freezing, and the snow was melting quite fast in the sun. The thermometer at the Lake Louise chalet reached seventy-two degrees at the same time that we were on the summit of Mount Temple, which proves this to be almost the highest temperature that ever occurs on this lofty point. It would be safe to say that the temperature on the top of Mount Temple never rises higher than forty degrees.

If one is fortunate in a good selection of routes, the ascent of Mount Temple will not be found difficult. But the descent is very perplexing, for unless one remembers the intricate combination of gullies and ledges by which the ascent is made, many precipitous cliffs will be encountered down which it is impossible to descend.

This was our last exploit in Paradise Valley, and a few days later the various members of our party, one by one, bade farewell to the beautiful region of Lake Louise with its many pleasant associations.

I remained there five or six weeks longer until winter commenced in earnest and drove every one away. During the first week of October I made a final visit to Paradise Valley with Mr. Astley, the manager of the chalet, in order to bring back our tent and the camping utensils. Snow covered the ground in the shady parts of the woods, even at the entrance of the valley. The stream had fallen so much that its rocky bed proved the best route up the valley, especially for the horse. After an hour's journey within the entrance we found ourselves at the base of Mount Sheol, and not far above us could be seen a fine herd of seven or eight mountain goats. They scampered off on seeing us, but soon came to halt as they were tempted by curiosity to have another look. These snow-white goats are the most characteristic animals of the Rockies and nearly correspond in habits with the more cunning chamois of Switzerland. Like them it is a species of antelope, though it resembles a goat to a remarkable degree.

We found our camp buried in snow, the ridge-pole of the tent broken down with the heavy burden, and everything so much disguised by the wintry mantle that we had difficulty in finding the camping place. Even as we were packing up the frozen canvas and blankets, the air was full of falling snow and the mountains encircling the valley were only revealed in vague and indefinite outlines, while ever and anon could be heard the dull roar of snow-slides sweeping down to the glacier.

About nightfall we were back at the entrance to the valley, where the lower altitude gave us the advantage of a ground nearly free of snow, though a fine rain sifted down through the spruce needles almost constantly.

Here we camped in the dense forest, and our roaring fire, built high with great logs, soon drove away the chill and dampness of the rainy night. The tent, our clothes, and the mossy ground were soon steaming, and the bright glare of our camp-fire illumined the trees and gave us good cheer, surrounded as we were by miles of trackless forests in the blackness of night. A hearty supper and a great pail of strong hot tea soon revived our spirits, and on a soft couch of heaths and balsam boughs—more luxurious than any bed of down—we bid defiance to the darkness and storm in perfect comfort. The next day the snow-flakes were falling gently and steadily, so that the trees were covered even to their branchlets and needles with the white mantle. The bushes, the mosses, and even the blades of grass in the swampy marshes, as we pursued our homeward way, were all concealed and transformed into pure white images of themselves in snow.

A few days later I went up to Lake Agnes to hunt for mountain goats, which frequent this place in great numbers. The snow was two feet deep. The lake was already nearly covered with ice, and I was compelled to seek shelter behind a cliff against a bitterly cold wind, driving icy particles of hail and snow against my face.

It was useless to prolong the contest longer. Winter had resumed her iron sway in these boreal regions and high altitudes, and in a few weeks Lake Louise too would begin to freeze, and no longer present its endless change of ripple and calm, light and shadow, or the reflected images of rocks and trees and distant mountains.

CHAPTER VIII.

The Selkirks—Geographical Position of the Range—Good Cheer of the Glacier House—Charming Situation—Comparison between the Selkirks and Rockies—Early Mountain Ascents—Density of the Forest—Ascent of Eagle Peak—A Magnificent Panorama—A Descent in the Darkness—Account of a Terrible Experience on Eagle Peak—Trails through the Forest—Future Popularity of the Selkirks—The Forest Primeval—An Epitome of Human Life—Age of Trees—Forests Dependent on Humidity.

WEST of that chain of the Rocky Mountains which forms the crest or backbone of the continent, lies another system of mountains called the Selkirk Range. Having many features in common with the mountains to the east, this range has, nevertheless, certain constant characteristics of vegetation and geological formation, so that the traveller who is but slightly familiar with them should never be at a loss in regard to his surroundings.

The position of this range in relation to the other mountains of the great Cordilleran System is not difficult to understand. The Selkirks may be said to begin in northwestern Montana between the Summit Range and the Bitter Root Mountains, and, trending in a northwestward direction through British Columbia about three hundred miles, they approach the main range and apparently merge into it near the Athabasca Pass. The most remarkable feature of the range is the manner in which it compels the great Columbia River to run northward for fifty leagues on its eastern side, before it allows a passage to the west, so that the northern portions of the range are entirely hemmed in by this large river, flowing in opposite directions on either side. Another feature of great interest in regard to the drainage is the relation between the Columbia and Kootanie rivers. The latter river is one of the chief tributaries to the upper Columbia, and flows southward to a point one mile and a half from the head waters of the Columbia, which it passes on its journey southward, while the Columbia flows in the opposite direction. The water of the Kootanie is actually higher than that of the Columbia at this point, and as the two rivers are only separated by a low, level plain, it was once proposed to cut a channel between, and divert the Kootanie into the Columbia.

GLACIER HOUSE.

The traveller is always glad to find himself at the

The Glacier House.

Glacier House in the heart of the Selkirks. This is more especially true, if in previous years, he has visited this charming spot and become in some degree familiar with the place. The railroad makes a large loop round a narrow valley and sweeps apparently close to the great glacier of the Selkirks, a vast sea of ice that glistens in a silvery white sheen and appears to rise above the forests as one looks southward. There is something pre-eminently comfortable and homelike about the Glacier House. The effect is indefinable, and one hardly knows whether the general style of an English inn, or the genuine hospitality that one receives, is the chief cause. One always feels at home in this wild little spot, and scarcely realizes that civilization is so far distant.

The rush of summer guests called for the erection of an annex, so that there are now two hotels for the accommodation of tourists. The Glacier House is located near the railroad, and occupies a small, nearly level, place at the bottom of one of those deep and narrow valleys characteristic of the Selkirks. Those who have visited the Franconia Notch in the White Mountains would be somewhat reminded of that beautiful spot upon first seeing the surroundings of Glacier. The ground in front of the hotel has been levelled and is rendered beautiful by a thick carpet of turf. In summer it is fragrant and almost snowy in appearance from the multitude of white clover blossoms. This garden spot in the wilderness is still further adorned by fountains, which break the continuity of the greensward, and are fed by cascades that may be seen descending the opposite mountain side in many a leap, through a total distance of 1800 feet.

But this small area, that man has improved and rendered more suitable to his comfort, is surrounded on all sides by a wilderness, perhaps better described as a little explored range of mountains separated by deep gorges and covered with dense forests. It is like the Alps of Switzerland and the Black Forest combined. There are snow-clad peaks, large glaciers, and *névé* regions of vast extent in the higher altitudes, while the valleys below are dark and sombre in their covering of deep, cool forests. The main range of the Rockies presents no such rankness of vegetable growth—mosses, ferns, and lichens covering every available surface on tree trunks and boulders—nor such huge trees as those found everywhere in the Selkirks.

Moreover, the mountains of the Selkirk Range probably average 1000 feet lower than in the corresponding parts of the main range, but nevertheless they seem white and brilliant in their mantles of everlasting snow and sparkling glaciers. Finally, one observes that the railroad track is covered at frequent intervals by snow-sheds of considerable length, constructed of heavy beams and massive timbers, in order to withstand the terrible force and weight of winter snow-slides and avalanches. In the main range of the Rockies there are no snow sheds. The question naturally arises—What is the reason of all these differences from the more eastern ranges?

The answer to the question is that the climate is more humid. The snowfall in winter is so great that it remains all summer at much lower altitudes than in the Rockies, and supplies glaciers, which descend perhaps a thousand feet nearer to sea-level. The moisture from this deep covering of snow, saturates the ground as it melts in the spring, and, in addition to frequent, heavy summer rains, nourishes the rich forests of these mountains. Moreover, the atmosphere is always slightly moister than it is to the east, and does not tend to dry up the ground or evaporate the mountain snows so rapidly as in the summit range.

The eastward movement of the atmosphere, carrying up moisture from the Pacific, causes a great condensation of clouds and a heavy rainfall as the air currents pass over the Selkirks, and leaves the atmosphere robbed of a great part of its moisture to pass over the next range to the east.

Almost all the differences between the Selkirks and the Rockies proper, spring from the single cause of a moister climate. The principal features of extensive snow fields and luxuriant forests can be readily understood. May not the deep, narrow valleys of the Selkirks be likewise explained from the more rapid action and greater erosive power of the mountain streams in cutting down their channels?

Whatever may be the cause of all these phenomena, the results are very apparent. Any one who has visited the Selkirks for an extended period has, without doubt, spent many a day within doors writing his diary or enjoying the pleasure of music or literature, while the rain is falling constantly, and the clouds and vapors hang low on the mountain sides. The manner in which the clouds come sweeping up the Illicellewaet valley at the base of Mount Cheops and turn toward the flanks of Eagle Peak or Mount Sir Donald is very impressive. Certainly the cloud effects in the Selkirks are magnificent beyond all description.

Nevertheless, it is not encouraging to have a friend step off the train and announce the fact that he has been enjoying fine weather for several days in the Rocky Mountains, some fifty or sixty miles to the east, while you have been confined to the house by a long period of rain.

Often, too, the climber or explorer becomes fretful under long confinement, and, taking advantage of an apparent clearing away of clouds and a promise of fair weather, when far from the hotel, is caught in a sudden downpour, and realizes the truth of that scriptural passage which was apparently written concerning a similar region —" They are wet with the showers of the mountains, and embrace the rock for want of a shelter."

When the railroad first made this region accessible to tourists, the Selkirks rapidly acquired a remarkable popularity, especially among mountain climbers. In this early period several parties came over from England and other countries of Europe with the express purpose of making mountain ascents. Such parties were those of Dr. Green and the two Swiss climbers, Huber and Sulzer. A good idea of the difficulties presented by the higher peaks to skilled mountaineers may be had from the fact that Dr. Green and his party only succeeded in reaching the summit of one high peak, while Huber and Sulzer left the Hermit Range in defeat, though they succeeded in reaching the top of the sharp rock peak, Mount Sir Donald, the Matterhorn of the Selkirks.

One of the chief difficulties to overcome is the penetration of the forest belt below the tree line. No one who has not tried a Selkirk forest has any conception of its nature in this respect. There are huge tree trunks lying on or near the ground, which have been thrown down by the precipitate fury of some winter snow slide, or have fallen by the natural processes of death and decay. These great obstacles are ofttimes covered with a slippery coating of moss and lichens, while the ground is fairly concealed by a rank growth of ferns, and plants in countless variety. The density of the underbrush is rendered still more trying to the mountaineer by reason of a plant of the Ginseng family, which from its terrible nature is most fitly named the Devil's Club, for it is armed with thousands of long needle-like spines. This plant grows five or six feet high, with a stout stem bearing a few leaves of large size. The spines, which are an inch or more in length, project in every direction like an array of quills on a porcupine, and are strong enough to penetrate the skin and flesh with surprising facility. The alder bushes attain a peculiar growth in the Selkirks; each bush consists of a bunch of long slender stems, which spread out from the ground in every direction, ofttimes with nearly prostrate branches, which interlace and form a wellnigh impassable hedge. The alder bushes are found most numerous on bare slopes of the mountains, where snow slides have stripped down the forests; or in ravines, where the crumbling earth gives no certain foothold to larger and nobler trees.

In 1893, A. and I made an ascent of Eagle Peak. This mountain lies just to the west from the great wedge-shaped rock summit of Mount Sir Donald. The altitude of Eagle Peak is, I believe, a little more than 9400 feet above sea-level, and as the Glacier House is only 4400 feet, the ascent involves a climb of 5000 feet. The name of the mountain is derived from a great crag or cliff near the summit, which appears to lean out from a ridge, and bears a striking resemblance to the head of an eagle. When we were making our ascent we came suddenly on the Eagle itself, which now, on a nearer view, proved to be of colossal size, a great leaning tower, about sixty feet high. Rising from one of the rocky ridges, it reached upwards and outwards till the outermost point seemed to overhang a bottomless abyss, perhaps twenty or twenty-five feet beyond the verge of the precipice.

The ridge just below the summit is a scene of wild confusion, for the rocky ledges have been split up and wedged apart by frost and storms till they appear as giant blocks of stone ten or fifteen feet high, between the crevices of which one may catch glimpses of the valley and forests thousands of feet below.

The view from the summit of Eagle Peak is magnificent and well worth the labor of the climb. The proximity of Mount Sir Donald, which towers more than 1200 feet higher, causes its sullen precipices to appear strikingly grand. The great Illicellewaet *névé*, with its twenty square

miles or more of unbroken snow fields, stretches out in the distance and forms part of the eastern horizon. The rugged appearance of the Hermit Range to the west, with its sharp ridges and needles, is perhaps the most tumultuous part in all this wild sea of mountain peaks. It has been stated on good authority that from Mount Abbott, a far lower ridge on the farther side of the valley, more than one hundred and twenty individual glaciers may be counted, but there are even more within view from Eagle Peak.

We remained on the summit till nearly three o'clock, and thereby took a great risk, as we learned afterwards to our exceeding regret. Before leaving, however, we built a high cairn and fixed several handkerchiefs among the stones so as to render it, if possible, visible from the valley below.

In our descent we found no trouble till we reached tree line, when the gathering gloom of nightfall, made earlier by a cloudy sky, aroused our apprehensions and led us to a serious mistake. Thinking that it would be better to follow the course of a stream, which had cut out a deep ravine in the mountain side, as there would be more light, for a time at least, we commenced our descent with all speed. We soon found ourselves in a trap, as the sides of the ravine grew constantly deeper and steeper as we descended, and it was at length impossible to get out at all. Floundering about among the long trailing branches of alders, our descent soon became a mixture of sliding, falling, and, indeed, every method of progress save rational walking. The darkness came on rapidly, as the days were short and the twilight much curtailed, it being late in the summer. In an hour it became so absolutely black that the foamy course of the stream we followed was the only visible object, as even the stars were concealed and their light shut out by a heavy covering of dark cloud. Sometimes the long, prostrate branches of the alders would catch our feet in a most exasperating manner, and cause one or the other to slide temporarily head-foremost, till some branch or root could be seized in the hand and the progress arrested. Once I saw a white object, just below me apparently, and thinking it might be a stone, was about to lower myself in fancied security when suddenly I realized that it was the foam of the stream some fifty feet below, and that we were on the edge of a precipice ! At another time I fell headlong through a bush and brought up against some great obstacle around which I wound my leg, not knowing whether it might be a huge grizzly or some other denizen of the forest, when sure enough it moved away, and rolled over my leg. It was a great boulder nearly a yard in diameter.

This nocturnal descent was the most bitter experience I have ever had in mountain climbing, as the anxiety and worry consequent upon each movement were exquisitely painful, and continued three hours. Arrived at the bottom of the slope at ten o'clock P.M., we found ourselves in the mass of fallen logs and debris near the stream, and likewise near the trail. Under the spell of a certain assurance that a few minutes more of toil would bring us out to the trail, we thought nothing of falling into holes

A Terrible Experience.

four or five feet deep, as we plunged about among the logs, or, when walking on them, occasionally stepped off into space.

We arrived at the Glacier House at 10:30 P.M., where we were surrounded by anxious friends, and regaled by a hot dinner of roasted chickens and all manner of good things, such as one always finds at this most excellent inn. At such times, more than at any other, one appreciates the thoughtfulness and care of a kind host.

Our experience on Eagle Peak, trying as it was, could not equal that of two gentlemen who, in 1894, made an attempt to scale the mountain. Unfortunately they failed to reach the summit, and, worse still, were benighted among the crags and cliffs at a high altitude, where they spent the night in misery. Finding themselves in their attempt unable to advance farther for some reason or other, they were descending, when it suddenly occurred to them that they were on a different ledge from any they had seen hitherto. Nightfall was bringing rapidly increasing darkness, and it seemed impossible, at length, either to proceed farther or even to retrace the steps by which they had come. Here, then, on a narrow ledge overlooking a precipice, the awful depths of which were rendered still more terrible in the obscurity of gathering gloom, and with their feet dangling over the verge, they were forced to remain motionless, and wear out the long night in cold and sleepless suffering. The next morning a search party was organized, and they were conducted back to the comforts of the Glacier House, much to the relief of their anxious friends, but nearly prostrated by their terrible experience.

Later, we made an ascent of Mount Cheops, a striking peak with a most perfect representation of a pyramid forming its summit. The view is fine but not worth the labor of the climb, as the ascent of the lower slopes seems interminably long and tedious by reason of the underbrush and steep slope. Like Eagle Peak, the summit revealed no evidence of previous conquests, and it will probably be a long time before any one will be so far led astray as to make a similar attempt.

Trails and good foot-paths lead from the Glacier House to points of interest in the vicinity. The chief resort is the Great Glacier itself, where one may witness all the phenomena of a large ice stream, or ascend to the vast *névé*, and wander about on a nearly level, and apparently limitless, snow field.

Mount Abbott is an easy and favorite climb, and is often successfully attempted by women who are endowed with considerable strength and endurance. On the way, a small pool, called Marion Lake, is passed. It nestles among the cliffs and forests on the mountain side far above the valley. It is the only lake I know of in the Selkirks. This is one of the remarkable differences between the Selkirks and the Summit Range of the Rockies: the absence of lakes in one region, and their great number in the other. The great majority of lakes in the Rockies are very small and often do not deserve the name, as they are mere pools a few yards across. But their small size in

Future of the Selkirks.   131

no way detracts from their beauty, and it is most unfortunate that the Selkirks possess so few of these, the most charming of all features in mountain landscapes.

The Selkirks are but little known, because the dense forests and the immense size of the fallen logs forbid the use of horses almost altogether, and will ever prevent the mountaineer from making extended journeys into the lesser known parts of the mountains, unless trails are cut and kept in good order. At present all provisions, blankets, and tents must be packed on men's backs, a method that is both laborious and expensive.

It must eventually result, however, that these mountains will prove a most popular resort for climbers and sportsmen. The attractions for either class are very great. For the mountaineer, they present all the grandeur and beauty of the Swiss Alps, with difficulties of snow and rock climbing sufficient to add zest to the sport. The multitude of unclimbed peaks likewise offers great opportunities for those ambitious for new conquests. The immense annual snowfall causes many of the higher peaks to assume an appearance of dazzling beauty and brilliancy, while the Alpine splendor of these higher altitudes is strongly contrasted with the dark-green color of the forested valleys.

For the sportsmen, too, there are abundant opportunities to hunt the larger game. On the mountains are numerous herds of mountain goats and sheep, while the forests abound in bears—the black bear and the grizzly or silver tip. During the berry season, these animals frequent the valleys and are often seen by the railroad men even near the Glacier House. One gentleman had the good fortune to shoot a black bear from a window of the hotel last year. Of course, there is practically no danger from even the grizzly bear in this immediate vicinity, as they have learned to fear man from being frequently shot at, and have long since lost the ferocity which they sometimes show in extremely wild and unfrequented regions.

No mention has yet been made of the kind of trees to be found in a Selkirk forest. Almost all the varieties of coniferous trees observed in the Rockies, except the Lyall's larch, occur in the Selkirks, though each variety attains much larger size. The cedar, the hemlock, the Douglas fir, and the Engelmann's spruce are most conspicuous and form the chief part of the forest trees. Each of these species here attains a diameter of from three feet upward, even to six or seven, and a height of from 150 to 200 feet.

Nothing is more enjoyable than to take one of the mountain trails and enter the depths of the forest, there to rest in quiet contemplation where trees alone are visible in the limited circle of view. On a quiet afternoon, when all is calm and not a breath of air is stirring, the long, gray moss hangs in pendent tufts from the lower branches of the giant trees, and one feels that this is indeed another Acadian forest of which Longfellow sings:

"This is the forest primeval. The murmuring pines and the hemlocks,
  Bearded with moss and in garments green, indistinct in the twilight,

The Forest Primeval.

Stand like Druids of eld, with voices sad and prophetic,—
Stand like harpers hoar, with beards that rest on their bosoms."

Such indeed is a Selkirk forest.

The idea that is at length developed in the mind, by a long rest in one of these deep and sombre forests, is that of the majesty, and silent, motionless power of vegetation. The creations of the vegetable world stand on all sides. They wellnigh cover the ground; they limit the horizon, and conceal the sky. The tall cedars have a shreddy bark that hangs in long strips on their tapering boles and makes the strongest contrast with the rough bark of the firs. What could be more unlike, too, among evergreens, than the spreading fanlike foliage of the cedars, the needle-like leaves of the firs, and the delicate spray of the hemlocks?

What a vast amount of energy has been preserved in these forest giants; with what a crash they would fall to the ground; and what a quantity of heat—which they have stored up from the sun through hundreds of summers—would they give out when burned slowly in a fireplace! If we examine a single needle, or a thin shaving of wood, under the microscope, and obtain a glimpse of the complexity of the cells and pores with which this vegetable life is carried on; or consider the wonderful processes by which the flowers are fertilized, and the cones mature, so that the species may never die out; and then regard the immensity of the whole forest stretching boundless in every direction, all constructed from an infinity of atoms, the mind and imagination are soon led beyond their depth.

' Now let the pure, cold light of science, with its precise and exact laws, fade away into the warm, mellow glow of romance, till we picture the forest as an epitome of human life, with its struggles, its suffering, and the slow but certain progress from infancy to old age and death. For here, among the forest trees, are every age and condition represented. Beneath, are young trees, vigorous and full of promise, hoping, as it were, some day to push their highest branches above the general plane of tree tops and share the life-giving sun, though, during the struggle, many will surely weaken and die in the pale and inefficient light beneath the older trees. Then there are the larger trees in the full glory of their prime, with massive trunks, straight and tall, giving promise of many years of life yet to come; and finally, the giants of the forest, their branches torn off by storms or their trunks rent and scarred by lightning. Everything about the oldest trees betokens the slow decay and all-conquering death, which is gradually sapping their life blood and pointing to their certain, final destruction. The long, gray moss, gently waving in the faintest breath of air, hangs from every limb, and makes these venerable monarchs resemble bearded patriarchs, which have stood here perhaps a thousand years battling with the elements, the wind, and the lightning, silent witnesses to the relentless progress of the seasons.

Trees have, however, all the qualifications of living forever. There is no reason why a tree should ever die, were it not for some unnatural cause, such as the fury of a storm, the rending power of lightning, or the destructive influence of insects and parasites. In California, in the Mariposa Grove, some of the giant redwood trees are twenty-five hundred years old. They began to grow when Solon was making laws for the ancient Greeks. These wonderful groves of California are, however, exceptional, and have survived by reason of the clemency of the climate and the fact that the aromatic redwood is avoided by insects. In most forests, the laws of chance and probability rarely allow the sturdiest trees to run the gamut of more than a few hundred years, and if they attain a thousand years, it is their "fourscore—by reason of strength."

In the Selkirks, one sees the ground covered with huge tree trunks in all stages of decay, slowly moldering away into a newer and richer soil; some have yielded to the natural processes of decay, others to accident or forest fires, while in some places winter avalanches have cut off the tops of the trees forty or fifty feet above the ground, and left nothing but a maze of tall stumps where once stood a noble forest.

The Selkirk forests are dense and sometimes almost magnificent in their luxuriance, and vastly surpass the forests of the eastern range in the variety of species, the size of the trees, and the luxuriant rankness of vegetable growth. At the same time they do not approach the almost tropical vigor and grandeur of the Pacific Coast forests, where a green carpet of moss covers the trunks and branches of the huge trees, and even ferns find nourishment in this rich covering, aided by the reeking, humid atmosphere, on branches forty or fifty feet above the ground. In such a forest, the ferns and brakes reach a height of six or eight feet above the ground, the various mosses attain a remarkable development, and hang in long, green tresses, a yard in length, from every branch, and exaggerate the size of the smaller branches, while the beautiful tufts of the Hypnum mosses appear like the fronds of small ferns, so large do they become.

The forests of the Summit Range, the Selkirks, and the Pacific Coast are almost perfect indexes of the humidity of the climate. The Selkirk forests are less vigorous than those of the Pacific coast, but more so than the light and comparatively open forests of the Summit Range, where the climate is much drier.

CHAPTER IX.

Mount Assiniboine — Preparations for Visiting it — Camp at Heely's Creek — Crossing the Simpson Pass — Shoot a Pack-Horse — A Delightful Camp — A Difficult Snow Pass — Burnt Timber — Nature Sounds — Discovery of a Beautiful Lake — Inspiring View of Mount Assiniboine — Our Camp at the Base of the Mountain — Summer Snow-Storms — Inaccessibility of Mount Assiniboine.

GREAT interest was aroused among tourists in the summer of 1895, by the reports of a remarkable peak south of Banff named Mount Assiniboine. According to current accounts, it was the highest mountain so far discovered between the International boundary and the region of Mounts Brown and Hooker. Besides its great altitude, it was said to be exceedingly steep on all sides, and surrounded by charming valleys dotted with beautiful lakes. The time required to reach the mountain with a camping outfit and pack-horses was said to be from five to seven days.

The romance of visiting this wild and interesting region, hitherto but little explored, decided me to use one month of the summer season in this manner. By great good fortune I met, at Banff, two gentlemen likewise bent on visiting the same region, and on comparing our prospective plans, it appeared that mutual advantage would be gained by joining our forces. In this way we would have the pleasure of a larger company, and at the same time the opportunity of separating, should we come to a disagreement.

The sixth of July was decided on as the date for our departure. In the meantime, we made frequent visits to the log-house of our outfitter, Tom Wilson, who was to supply us with horses, our entire camping outfit, and guides. Many years previously, Wilson had packed for the early railroad surveyors, and had thus gained a valuable experience in all that concerns the management and care of pack-animals among the difficulties of mountain trails. In the past few years, he has been engaged in supplying tourists with camping outfits and guides, for excursions among the mountains.

The season of 1895 was very backward, and there was an unusually late fall of snow at Banff, in the middle of June. Moreover, the weather had remained so cold that the snow on the higher passes still remained very deep, and several bands of Indians, who attempted to cross the mountains with their horses late in June, were repulsed by snow six or eight feet deep.

The weather continued cold and changeable during the first week in July. In the meanwhile, however, our preparations for departure went on without interruption, and Wilson's log-house, where the supplies and camp outfits were safely stored, became a scene of busy preparation.

On every side were to be seen the various necessaries of camp life: saddles for the horses, piles of blankets, here and there ropes, tents, and hobbles. Great heaps of provisions were likewise piled up in apparent confusion, though, in reality, every item was portioned out and carefully calculated. Rashers of bacon and bags of flour comprised the main bulk of the provisions, but there were, besides, the luxuries of tea, coffee, and sugar, in addition to large quantities of hard tack, dried fruits and raisins, oatmeal, and cans of condensed milk. Pots and pails, knives, forks, and spoons, and the necessary cooking utensils were collected in other places. Our men were already engaged for the trip, and were now busily moving about, seeing that everything was in order, the saddle girths, hobbles, and ropes in good condition, the axes sharp, and the rifles bright and clean.

At length the sixth of July came, but proved showery and wet like many preceding days. Nevertheless, our men started in the morning for the first camp, which was to be at Heely's Creek, about six miles from Banff. Our prospective route to Mount Assiniboine was, first, over the Simpson Pass to the Simpson River, and thence, by some rather uncertain passes, eastward, toward the region of the mountain.

Toward the middle of the afternoon we started on foot for Heely's Creek, where our men were to meet us and have the camp prepared. Passing northward up the valley, we followed the road by the famous Cave and Basin, where the hot sulphur water bubbles up among the limestone formations which they have deposited round their borders. The Cave appears to be the cone or crater of some extinct geyser, and now a passage-way has been cut under one wall, so that bathers may enjoy hot baths in this cavern. A single opening in the roof admits the light.

A short time after leaving these interesting places, we had to branch off from the road, and plunge into a burnt forest, where there was supposed to be a trail. The trail soon faded away into obscurity among the maze of logs, and, worse still, it now came on to rain gently but constantly. After an hour or more of hard work we came to Heely's Creek.

The camp was on the farther side of the creek, and, after shouting several times, Peyto, our chief packer, came dashing down on horseback, and conveyed us, one at a time, across the deep, swift stream. Peyto made an ideal picture of the wild west, mounted as he was on an Indian

PEYTO.

steed, with Mexican stirrups. A great sombrero hat pushed to one side, a buckskin shirt ornate with Indian fringes on sleeves and seams, and cartridge belt holding a hunting knife and a six-shooter, recalled the romantic days of old when this was the costume throughout the entire west.

Our encampment consisted of three tents, prettily grouped among some large spruce trees. A log fire was burning before each tent, and, on our arrival, the cooks began to prepare our supper. This was my first night in a tent for a year, and the conditions were unfavorable for comfort, as we were all soaked through by our long tramp in the bush, and, moreover, it was still raining. Nevertheless, we were all contented and happy, our clothes soon dried before the camp fires, and after supper we sang a few popular songs, then rolled up in warm blankets on beds of balsam boughs, and slept peacefully till morning.

I was awakened at dawn by the cry of "Breakfast is ready," and prepared forthwith to do it justice. The day appeared cloudy but not very threatening. In an hour the packers began their work, and it was wonderful to observe the system and rapidity of their movements. The horses, of which we had seven as pack-animals and two for the saddle, were caught and led to the camp, where they were tied to trees near by. All the provisions, tents, cook boxes, bags, and camp paraphernalia were then made ready for packing. There are three prime requisites in skilful packing. They are: the proper adjustment of the blanket and saddle so that it will neither chafe the back of the horse nor slip while on the march; the exact balancing of the two packs; and the knowledge of the

PACKING THE BUCKSKIN.

"diamond hitch." The wonderful combination of turns and loops which go to make up the diamond hitch has always been surrounded with a certain secrecy, and jealously guarded by those initiated into the mysteries of its formation. It was formerly so essential a part of the education of a Westerner that as much as one hundred dollars have been paid for the privilege of learning it. Without going into details, it may be described as a certain manner of placing the ropes round the packs, which, once learned, is exceedingly simple to tie on or take off, and it will hold the pack in place under the most trying circumstances. The name is derived from a diamond-shaped figure formed by the ropes between the packs.

By eight o'clock our procession of ten horses was on the march, and, after passing through a meadow where every blade of grass was hung with pendent drops of mingled rain and dew, now sparkling bright in the morning sun, we came to the trail. Our winding cavalcade followed near the creek and gradually rose above its roaring waters, which dashed madly over many a cascade and waterfall in its rocky course. Our pathway rose constantly and led us through rich forests.

Peyto led the procession mounted on an Indian horse called Chiniquy, not a very noble-looking beast, but a veteran on the trail, and, by reason of his long legs, a most trustworthy animal in crossing deep rivers. Then followed the pack-horses with the men interspersed to take care of them, and the rear was brought up by our second packer, likewise on horseback. The greater part of the time, the gentlemen of the expedition kept in the rear.

The flowers were in all the glory of their spring-time luxuriance, and we discovered new varieties in every meadow, swamp, and grove. Beside the several varieties of anemones, the yellow columbines, violets, and countless other herbaceous plants, we found, during the march of this day, six kinds of orchids. Among them was the small and beautiful, purple Calypso, which we found in bogs and damp woods, rearing its showy blossom a few inches above the ground.

CALYPSO.

At the base is a single heart-shaped leaf. We were very much pleased to find this elegant and rare orchid growing so abundantly here. There is a certain regal nobility and elegance pertaining to the whole family of orchids, which elevates them above all plants, and places them nearest to animate creation. Whether we find them in high northern latitudes, in cold bogs, or in dark forests, retreating far from the haunts of men, avoiding even their own kind, solitary and unseen; or perhaps crowded on the branches of trees in a tropical forest, guarded from man by venomous serpents, the stealthy jaguar, stinging insects and a fever-laden air; they command the greatest interest of the botanist and the highest prices of the connoisseur.

We camped at about two o'clock, not far from the summit of the Simpson Pass, in a valley guarded on both sides by continuous mountains of great height.

We were surprised the next day, on reaching the summit, to find the pass covered with snow, heaped in great drifts, ten or twenty feet deep, among the trees. The Simpson Pass is only 6884 feet above tide, and, consequently, is below the tree line. Near the summit were two small ponds still frozen over. A warm sun and a genial south wind were, however, rapidly dissolving the snow and reducing it to slush, while clear streams of water were running in the meadows everywhere, regardless of regular channels.

As we began our descent on the south side, a great change came over the scene. Two hundred feet of descent brought us from this snowy landscape to warm mountain slopes, where the grass was almost concealed by reason of myriads of yellow lilies in full blossom, mingled with white anemones. These banks of flowers, resembling the artificial creations of a hot-house, were sometimes surrounded on all sides by lingering patches of snow. Such constant and sudden change is characteristic of mountain climates, where a few warm days suffice to melt the snow and coax forth the flowers with surprising rapidity.

The trail now descended rapidly, and led us through forests much denser and more luxuriant than those on the other side of the pass. Everything betokened a moister climate, and the character of the vegetation had changed so much that many new kinds of plants appeared, while those with which we were familiar grew ranker and larger. We had crossed the continental divide, from Alberta into British Columbia.

Early in the afternoon we came to our camping place on the banks of the Simpson River, where a great number of teepee poles proved this to be a favorite resort among the Indians. On all sides, the mountains were heavily forested to a great height, and, far above, gray limestone cliffs rose in bare precipices nearly free of snow.

On July the ninth, we made the longest and most arduous march so far taken. Our route, at first, lay down the Simpson River for several miles. While the horses and men followed the river bed almost constantly, making frequent crossings to avail themselves of better walking and short cuts, the rest of us necessarily remained on one bank, and were compelled to make rapid progress to keep up with our heavily laden horses.

After we had proceeded down the winding banks of the Simpson River for about two hours, our pass, a mere notch in the mountains, was descried by Mr. B., who had visited this region two years before in company with Wilson. The pass lay to the east, and it was necessary for every one to cross the river, which was here a very swift stream nearly a yard in depth. We all got across in safety, but had not advanced into the forest on the farther side more than fifty yards, when one of my pack-horses fell, by reason of the rough ground, and broke a leg. It required but a few minutes to unpack the poor beast and end his career with a rifle bullet. The packs were then placed on old Chiniquy, the faithful beast hitherto used by Peyto as a saddle-horse.

In less than fifteen minutes we were ready to proceed again. The trail now led us up very steep ascents on a forest-clad mountain slope for several hours. After this we entered a gap in the mountains and followed a stream for many miles, and at length pitched our camp late in the afternoon, after having been on the march for nine hours.

Every one was rejoiced at the prospect of a rest and something to eat. Even the horses, so soon as their packs and saddles were removed, showed their pleasure by rolling on the ground before hastening off to a meadow near by. Axes were busy cutting tent poles and fire-wood. Soon the three tents were placed in position; and fires were burning brightly before each, while the cooks prepared dinner.

This place was most delightful. The immediate

ground was quite level and grassy. Near by was a clear deep stream with a gentle, nearly imperceptible current, which afforded a fine place for a cold plunge. The mountains hemmed in a valley of moderate width and presented a continuous barrier on either side for many miles. The general character of the scenery was like that of the Sierra Nevadas, with high cliffs partly adorned with trees and shrubs, down which countless waterfalls fell from heights so great, that they resembled threads of silver, waving from side to side in the changing currents of air. On the mountain side south of our camp, there stood a remarkable castle or fortress of rock, where nature had apparently indulged her fancy in copying the works of men. So perfect was the representation, that no aid from the imagination was required to see ramparts, embrasures, and turreted fortifications of a castle, in the remarkable pinnacles and clefts cut out by nature from the horizontal strata. The next morning, every one was more or less inspired with a pleasing anticipation and excitement, as, according to reports, we had not far to go before we should get our first view of Mount Assiniboine. At the end of our valley was a pass, from the summit of which Mount Assiniboine could be seen. The trail led us through a forest with but little underbrush, and presently a beautiful lake burst on our view. Two of us, being somewhat in advance of the pack train, caught a dozen fine trout here in a very short time, and were only interrupted by the arrival of the horses and men. The fish were so numerous that they could be seen everywhere on the bottom, and at the appearance of our artificial flies on the water, several fish would rise at once.

In half an hour, the summit of our pass appeared over the tree tops, and rose, apparently, 500 feet higher. The state of the pass was, however such as to cool our enthusiasm decidedly. It was completely covered with snow to a great depth, which made it seem probable that we would not succeed in getting the horses over. As this could not be proved from our position, we pushed on, determined to overcome all difficulties. The snow began to appear, at first, in small patches in shady places among the forest trees, then in large drifts and finally, everywhere except on the most exposed slopes. The trail had been lost for some time, buried deep in the snow. Our progress was not difficult, however, as the forest had assumed the thin and open nature characteristic of high altitudes, and it was possible to proceed in any direction. Our horses struggled on bravely, and by dint of placing all the men in front and breaking down a pathway, we managed to effect passages over long stretches where the snow was five or six feet deep. After the tree line had been reached, we were more fortunate, as a long narrow stretch, free of snow led quite to the top of the pass, through the otherwise unbroken snow fields. A great cornice of snow appeared on our right near the top of the pass and showed a depth of more than forty feet.

Near the top of the pass the travelling was much easier, and in a few minutes we were looking over the summit across a wide valley to a range of rough moun-

A Difficult Snow Pass.

tains hung with glaciers. Beyond them, and rising far above, could be seen the sharp crest of Mount Assiniboine, faintly outlined against the sky in a smoky atmosphere. The intervening wide valley revealed a great expanse of burnt forest. The dreary waste of burnt timber was only relieved by two lakes, several miles distant, resting in a notch among the mountains. The nearer was about a mile in length, while slightly beyond, and at a higher elevation, was the second, a mere pool

APPROACHING THE PASS.

of dark blue water, resting against the moraine of a glacier.

In the valley, a meadow near a large stream seemed to offer the best chances for a camp. In an hour we reached this spot after a hard descent. Some of our horses displayed great sagacity in selecting the safest and easiest passages between and around the logs, and gave evidence of their previous experience in this kind of work.

In order to rest the men and horses, after the arduous marches of the past forty-eight hours, we decided to remain an entire day at this place. We were also anxious to explore the two lakes, as they seemed to offer fair promise of beautiful scenery and interesting geological formation. Our camp was surrounded on all sides by burnt forests and charred logs, and so offered but little of the picturesque. A partial compensation was enjoyed, however, by reason of the great variety and number of song birds which were now nesting in a small swamp near by. This bog was clothed in a rich covering of grass, and here our horses revelled in the abundance of feed, while some small bushes scattered here and there afforded shelter and homes for several species of birds. All day long and even far into the night we were entertained by their melodies. The most persistent singer of all was the white-crested sparrow, whose sweet little air of six notes was repeated every half minute throughout the entire day, beginning with the first traces of dawn. Perhaps our attention was more attracted to the sounds about us because there was so little to interest the eye in this place. Smoke from distant forest fires obscured whatever there was in the way of mountain scenery, while the waste of burnt timber was most unattractive. A warm, soft wind blew constantly up the valley and made dull moanings and weird sounds among the dead trees, where strips of dried bark or splinters of wood vibrated in the breeze. The rushing stream, fifty yards from our camp, gave out a constant roar, now louder, now softer, according as the wind changed direction and carried the sound towards or away from us. The thunders of occasional avalanches, the loud reports of stones falling on the mountain sides, were mingled with the varied sounds of the wind, the rustling of the grass, the moaning trees, and the songs of birds. These were all pure nature sounds, most enjoyable and elevating. Though but partially appreciated at the time, such experiences linger in the memory and help make up the complex associations of pleasures whereby one is led to return again and again to the mountains, the forests, and the wilderness.

Our time, which was set aside for this region, now being consumed, we started on July the twelfth for the valley at the base of Mount Assiniboine, where it was probable that we should camp for a period of two weeks or more. Our route lay toward the end of the valley and thence around a projecting spur of the mountain which cut off our view. In about two hours our horses were struggling up the last steep slope near the summit of the divide. I had delayed for a photograph of a small lake, so the horses and men were ahead. When at length I gained the top I found that a misplaced pack had caused delay, and so I overtook the entire party on the borders of a most beautiful sheet of water. The transformation was nearly instantaneous. The burnt timber was completely shut out from view by the low ridge we had just passed over, and we entered once more a region of green forests. The lake was long and narrow ; on the farther side, hemmed in by rock slides and cliffs of the mountains, but on the west side a trail led along the winding shore among larch and spruce trees. In many shady nooks along the banks of the lake were snow-drifts, under the trees or behind protecting rocks. So long had winter lingered this season that part of the lake was still covered with ice. Large fragments of ice were drifting down the lake and breaking among the ripples. Near the shore in some places, the water was filled with thousands of narrow, needle-like pieces of ice several inches long and perhaps thick as a match, which, by their rubbing together in the moving water, made a gentle subdued murmur like the rustling of a silken gown. When ice is exposed to a bright sun for several days, it shows its internal structure by separating into vertical columns, with a grain like that of wood. The ice needles which we saw had been formed during the last stages of this wonderful process.

The Indians had made a most excellent trail round the lake, as frequently happens in an open country. Wherever dense brush or much fallen timber occur, the trail usually disappears altogether, only to be discovered again where there is less need for it. It is said that a trail, once made, will be preserved by the various game animals of the country. In fact, there were quite recent tracks of a mountain goat in the path we followed around the lake.

The trail closely followed the water's edge and led us to the extreme end of the lake and thence eastward, where, having left this beautiful sheet of water, we passed through a grove for a very short space and came at once to another smaller, and possibly still more beautiful, sheet of water. Simultaneously the magnificent and long-expected vision of Mount Assiniboine appeared. It was a most majestic spire or wedge of rock rising out of great snow fields, and resembling in a striking manner the Matterhorn of Switzerland.

It would be impossible to describe our feelings at this sight, which at length, after several days of severe marching, now suddenly burst upon our view. The shouts of our men, together with the excitement and pleasure depicted in every face, were sufficient evidence of our impressions. After a short pause, while we endeavored to estimate the height and distance and gain some true idea of the mountain, all moved on rapidly through alternating groves and meadows to our camping place. This was at length selected about a half mile from the place where we first saw Mount Assiniboine. Here was a lake nearly a mile long, which reached up nearly to the base of the mountain, from which it was separated by a glacier of considerable size. Our camp was on a terrace above the lake, near the edge of a forest. A small stream ran close to our tent, from which we could obtain water for drinking and cooking purposes. The lake was in the bottom of a wide valley, which extended northwards from our camp for several miles, and then opened into another valley running east and west. The whole place might be described as an open plain among mountains of gentle slope and moderate altitudes, grouped about Assiniboine and its immediate spurs.

Our camp was 7000 feet above sea-level, and this was the mean height of the valley in all this vicinity. On mountain slopes this would be about the upper limit of tree growth, but here, owing to the fact that the whole region was elevated, the mean temperature was slightly increased, and we found trees growing as high as 7400 or 7500 feet above sea-level. Nevertheless, the general character of the vegetation was sub-alpine. Many larches were mingled with the balsam and spruce trees in the groves, and extensive areas were destitute of trees altogether. These moors were clothed with a variety of bushy plants, mostly dwarfed by the rigor of the climate, while here and there a small balsam tree could be seen, stunted and deformed by its long contest for life, and bearing many dead branches among those still alive. These bleached and lifeless limbs, with their thick, twisted branches resisting the axe, or even the approach of a wood-cutter, resembled those weird and awful illustrations of Doré, where

evil spirits in the infernal regions are represented transformed to trees.

Curiously enough, the trees in the groves were more or less huddled together, as though for mutual protection. The outlying skirmishers of balsam or spruce were undersized, and often grew in natural hedges, so regular that not one single branchlet projected beyond the smooth surface, as if sensitive of the wind and cold. The vegetable world does not naturally excite our sympathy, but this exhibition of, as it were, a united resistance against the elements was almost pitiable.

Snow banks surrounded our camp and appeared everywhere in the valley. The lake was not entirely free of ice, and large pieces of snow and ice, dislodged from the shores, were drifting rapidly down the lake, driven on by a strong wind and large waves. The whole picture resembled a miniature Arctic sea, where the curiously formed pieces of ice, often T-shaped and arched over the water, recalled the characteristic forms of icebergs.

It was at first impossible to explain where this never-failing supply of ice came from. What was our surprise, on making an exploration of the lake, to find that it had no outlet and was rapidly rising! The snow banks and masses of ice near the glacier were being gradually lifted up and broken off by the rising water, and so floated down the lake.

We remained at Camp Assiniboine for two weeks. During this time we ascended many of the lesser peaks in the vicinity, and made excursions into the neighboring valleys on all sides. The smoke only lasted one day after our arrival, but, unfortunately, the weather during the first week was very uncertain and fickle. A succession of storms, very brief but often severe, swept over the mountains and treated us to a grand exhibition of cloud and storm effects on Mount Assiniboine. Sometimes the summit would be clear, and sharply outlined against the blue sky, but suddenly a mass of black clouds would advance from the west and envelope the peak in a dark covering. Long streamers of falling snow or rain would then approach, and in a few moments we would feel the effects at our camp. During these mountain storms the wind blows in furious gusts, the air is filled with snow or sometimes hailstones, while thunder and lightning continue for the space of about ten minutes. The clouds and storm rapidly pass over eastward, and the wind falls, while the sun warms the air, and in a few minutes removes every trace of hail or snow. Thus we were often treated to winter and summer weather, with all the gradations between, several times over in the space of an hour.

It seemed impossible to ascend Mount Assiniboine, guarded as it was by vertical cliffs and hanging glaciers. Only one route appeared on this side of the mountain, and this lay up the steep snow-covered slope of a glacier, guarded at the top by a long *schrund* and often swept by rocks from a moraine above. It might be possible, having gained the top of this, to traverse the great *névé* surrounding the rock peak of Mount Assiniboine. From the snow fields the bare rock cliffs rise about 3,000 feet. The angle of slope on either side is a little more than fifty-one degrees, a slope which is often called perpendicular, and, moreover, as the strata are horizontal, there are several vertical walls of rock, which sweep around the entire north and west faces, and apparently make impassable barriers.

NORTH LAKE—LOOKING NORTHWEST.

CHAPTER X.

Evidence of Game—Discovery of a Mountain Goat—A Long Hunt— A Critical Moment—A Terrible Fall—An Unpleasant Experience— Habitat of the Mountain Goat—A Change of Weather—A Magnificent Panorama—Set out to Explore the Mountain—Intense Heat of a Forest Fire—Struggling with Burnt Timber—A Mountain Bivouac—Hope and Despair—Success at Last—Short Rations—Topography of Mount Assiniboine—The Vermilion River—A Wonderful Canyon—Fording the Bow River.

DURING our excursions we met with but little game, though it was very evidently a region where wild animals were abundant. The ground in many places was torn up by bears, where they had dug out the gophers and marmots. Large pieces of turf, often a foot or eighteen inches square, together with great stones piled up and thrown about in confusion around these excavations, gave evidence of the strength of these powerful beasts.

Higher up on the mountains we saw numerous tracks of the mountain goat, and tufts of wool caught among the bushes as they had brushed by them.

I was strolling through the upper part of the valley late one afternoon, when my eye fell suddenly on a mountain goat walking along the cliffs about a quarter of a mile distant. I had no rifle at the time and so returned to camp for one, meanwhile keeping well covered by trees and rocks. In a quarter of an hour I was back again and saw the goat disappear behind a ledge of rock about a half mile distant. The mountain goat always runs up in case of danger, so that it is essential to get above them in order to hunt successfully. I started forthwith to climb to a ledge about 200 feet above the one on which the goat appeared. This involved an ascent of some 600 feet, as the strata had a gentle dip southward toward Mount Assiniboine, so that it was necessary to take the ledge at a higher point and follow the downward slope. I was well covered by intervening cliffs, and the wind was favorable. It seemed almost a certainty that I should get a shot by following this ledge for about a mile. Accordingly I moved rapidly at first, and afterwards more cautiously, expecting to see the goat at any moment. At length I came to a narrow gorge, partially filled with snow, where there were fresh tracks leading both up and down. On a further study of the problem, I saw fresh tracks in the snow of the valley bottom, and knowing that it would be nearly useless to go up for the goat, I took the alternative chance of finding the animal below. After a hunt of two hours I returned to camp completely baffled. Arrived there, I caught sight of the goat standing unconcernedly on a still higher ledge.

It was now late in the day, but after a good camp dinner I set off again, determined to have that goat if it was necessary to stalk him all night. The animal was resting on a ledge near the top of a precipice fully 250 feet in height. I studied his position for at least a quarter of an hour, carefully noting the snow patches on the ledge above, so that it would be easy to recognize them on arriving there. Having made sure that I could recognize the exact spot below which the goat was located, I started to climb, and by a rough estimate calculated that I should have to ascend at least 1000 feet. After a few hundred yards, I was completely hidden from the goat in a shallow gully. Urged on by the excitement of the hunt, I reached the ledge in twenty minutes and turned southward. I now had to scramble over and among some enormous blocks of stone which had fallen from the mountain side and were strewn about in wild disorder. Some were twenty feet high, and between them were patches of snow which often gave way very suddenly and plunged me into deep holes formed by the snow melting back from the rock surfaces. Very soon I came to a small pool of water and a trickling stream, already freezing in the chill night air.

It was after nine o'clock, though there was still a bright twilight in the northwest, somewhat shaded, however, by the dark cliffs above. I proceeded very slowly, so as to cool down somewhat and become a little steadier after the rapid ascent. In about ten minutes I recognized the patch of snow under which the goat was located, about one hundred yards ahead. I went to the edge of the precipice cautiously, with rifle ready, and examined the ledges below. The up-draught, caused by the sun during the daytime, just now changed to the downward flow of the night air, chilled by radiation on the mountain side. This I thought would arouse the goat, but just at that moment my foot slipped and I dislodged a few pieces of loose shingle, which went rattling down the cliffs. These stones made the goat apprehensive of danger, in all probability, for I had no sooner recovered my balance than I caught sight of the white head and shoulders of the animal about twenty-five yards below. The animal stood motionless and stared at me in a surprised but impudent manner. I took aim, but could not keep the sight on him long enough to make sure of a shot, as my rapid climb had made my nerves a trifle unsteady. Fortunately, the goat showed not the slightest disposition to move and in a few seconds I got a good aim and fired. As soon as the smoke cleared, I saw a dash of white disappearing, and then heard a dull thud far below. A few seconds later I saw the animal rolling over and over down the mountain side, where it finally stopped on a slide of loose stones. I had to make a long detour in order to get down to the animal, where I arrived in about half an hour, and, remarkably enough, both horns were uninjured, though the goat had fallen 125 feet before striking. This good luck resulted from a small snow patch at the base of the cliff, which had broken the force of the fall, and here there was a perfect impression of the animal's body, eighteen inches deep, in the hard snow, while the next place where he had struck was about fifteen feet below.

It was about 10:30 o'clock when I started for camp, and so dark, at this late hour, that it was just possible to distinguish the obscure forms of rocks and trees on the mountain side. There was still another ledge to be passed before I could get down to the valley, where the only recognizable landmarks were occasional snow patches, and a single bright gleam in the darkness—our camp fire. I traversed northwards in descending, so as to pass beyond the vertical ledge, and at length, thinking that I had gone far enough, tried to descend. The place was steep, but as there were a few bushes and trees a safe descent seemed practicable. So I unslung my rifle, and, after resting it securely in a depression, I lowered myself till my feet rested on a projection of rock below. At the next move there was great difficulty in finding a rest for the rifle. At length I found a fair place, and lowered myself again. One more step and I should reach the bottom. Fortunately there was a stout balsam tree at the top of the ledge, with great twisted roots above the rocks, which would afford excellent hand-holds. Grasping them, after placing the rifle in the lowest place, I lowered myself again, but to my surprise I could not touch the bottom, and, looking down, found that I was hanging over a ledge twenty feet high with rough stones below. Just then the rifle began to slip down, as in my movements I had disturbed some bushes supporting it. With one hand firmly grasping a stout root, and the toe of my boot resting against the cliff, I took the rifle in my other hand, and after a most tiresome struggle, succeeded at length in placing it secure for the moment. It was now a hand-over-hand contest to get up. In going down everything had seemed most firm and secure, but now it was impossible to rely on anything, as the bushes broke away in my hand or were pulled out by the roots, and the rocks all appeared loose or too smooth to grasp. Necessity, however, knows no law, and after a most desperate effort I regained the top of the cliff. Not relishing any more experiences of this nature, I groped my way along for some distance and finally found an easy descent. On reaching the valley, the snow patches here and there afforded safe routes, illumined, as they were, by the starlight. I reached camp after eleven o'clock tired but successful.

My men started at five o'clock in the morning with ropes and a pole to bring down the game. It was a fine young male, and we found the meat a most pleasing addition to our ordinary fare. Goat meat has always had a bad reputation among campers and explorers, by reason of its rank flavor. This, however, probably depends on the age and sex of the animal, or the season of year. In all those that I have tried there was merely a faintly sweet flavor, which, however, is not at all apparent if the meat is broiled or roasted, and it is then equal to very fair beef or mutton.

The mountain goat inhabits the cliffs and snowy peaks of the Rockies, from Alaska to Montana and Idaho, and thence southward in certain isolated localities. Both sexes are furnished with sharp black horns curving gracefully backwards. The muzzle and hoofs are jet black, but the wool is snow-white, long, and soft, making a beautiful rug if the animal is killed in winter. Then the hair becomes very long, and the soft thick wool underneath is so dense as to prevent the fingers passing through.

Though these strange animals resemble true goats to a remarkable degree, and the old males sometimes have beards in winter, they are really a species of antelope, closely related to the chamois of Switzerland. They do not resemble those animals in wariness and intelligence, but are rather stupid and slow in getting out of danger. They are, however, pugnacious, and, when brought to bay, will often charge on the hunter and work fearful damage with their sharp horns. The legs are exceedingly stout and so thickly covered with long hair as to give the animal a clumsy appearance. Their trails are almost always to be found traversing the mountain sides, far above the tree line, at the bases of cliffs, and often passing over the lowest depression into the next valley. These goat tracks are so well marked that they often help the mountaineer, and sometimes lead him over places where without their guidance it would be impossible to go. The gait of the animal when running is a sort of gallop, which appears rather slow, but when one considers the nature of the ground they traverse, it is very rapid. The most inaccessible cliffs, frozen snow fields, or crevassed glaciers offer no barriers to these surefooted animals. I have seen a herd of several goats bounding along on the face of the cliffs, where it did not appear from below that there could be any possible foothold.

When a herd of goats come to a gorge or passage of

any kind where loose stones are liable to be dislodged on those below, these skilful mountaineers adopt the same plan of progress practised by human climbers. While the herd remains below, under the protection of the cliffs, one goat climbs the gully, and upon arriving at the top another follows, and thus, one by one, all escape danger.

The mountain goat is difficult to hunt by reason of the amount of climbing necessary to get near them, or above them. They are far less wary than the chamois of Switzerland, or the Rocky Mountain sheep. Nevertheless, they seem to be endowed with a wonderful vitality, and are very hard to kill. A goat not fatally wounded will

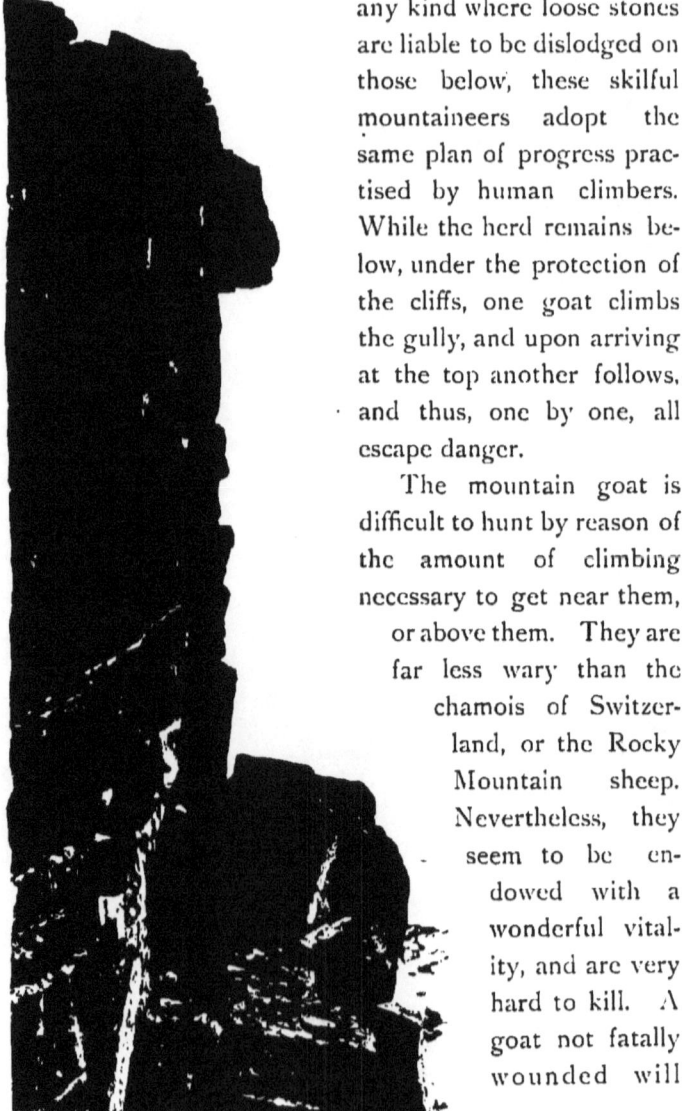

'HAUNT OF THE MOUNTAIN GOAT.'

often jump from a cliff on which he is standing, and survive a considerable fall. A friend of mine shot a goat near Lake Louise, which, after the first bullet, rolled down a cliff more than thirty feet high and landed on its feet at the bottom, where it proceeded to walk off as though nothing unusual had happened. The animal I shot near Mount Assiniboine fell 125 feet, and then rolled 200 feet farther, and was still alive when I reached him half an hour later.

These animals are by far the most numerous of the big game in the Canadian Rockies, and are said to be increasing in numbers. Their habits of frequenting high altitudes and inaccessible parts of mountains will tend to preserve them for many years from the relentless hunter.

After a week of fickle weather with five inches of new snow on July 15th, there was a decided change for the better, and the warm, bright days following one another more regularly gave us the first taste of real summer that we had. The massed drifts of snow diminished from day to day and the ice disappeared from the lakes. Nature, however, tempered her delights by ushering in vast numbers of mosquitoes and bull-dog flies to plague us. I was engaged at this time in some surveying work, in order to determine the height of Mount Assiniboine, and had to exercise the utmost patience in sighting the instruments, surrounded by hundreds of voracious foes, and often had to allow my face and hands to remain exposed to their stings for several minutes.

We obtained the most imposing view of Mount Assiniboine from the summit of a mountain about five miles

east of our camp. Standing at an altitude of 8800 feet, there were eighteen lakes, large and small, to be seen in the various valleys, which, together with the tumultuous ranges of the Rocky Mountains on every side, some of them fifty or sixty miles distant, formed a magnificent panorama.
From this point, which was nearly due north of Mount Assiniboine, the mountain shows an outline altogether different from that seen at our camp.

MOUNT ASSINIBOINE FROM NORTHWEST.

Here it forms a magnificent termination of a stupendous wall or ridge of rock, about 11,000 feet high, which runs eastward for several miles, and then curving around to the north, rises into another lofty peak nearly rivalling Mount Assiniboine in height. A very large glacier sweeps down from the *névé* on the north side of this lesser peak, and descends in a crevassed slope to the valley bottom.

The valley just east of us was quite filled by three lakes, the uppermost deep blue, the next greenish, and a smaller one, farther north, of a yellowish color.

Our last exploit at Mount Assiniboine was to walk

completely around the mountain. We had long desired to learn something of the east and south sides of this interesting peak, and to effect this Mr. B., Peyto, and I started on July 26th, determined to see as much as possible in a three days' trip. Our provisions consisted of bacon, hard tack, tea, sugar, and raisins. Besides this we carried one blanket apiece, a small hand axe, and a camera. As our success would depend in great measure on the rapidity of our movements, we did not burden ourselves with ice-axes or firearms except a six-shooter. After bidding farewell to Mr. P. and the other men in camp, and telling them to expect us back in three days, we left our camp at eight o'clock in the morning. We walked for three miles through the open valleys to the north and east, and in about two hours stood at the top of the pass, some 8000 feet above sea-level. From here we made a rapid descent for about 2000 feet, to the largest lake of this unexplored valley, which probably supplies one of the tributaries to the Spray River. The change in the character of the vegetation was remarkable. The trees grew to an immense size and reminded me strongly of a Selkirk forest. We had a most difficult scramble here in the pathless forest and up the opposite side of the valley. The heat was oppressive, and we were glad to gain the level of another more elevated valley where a cooler atmosphere greeted us. We held our way eastward for several miles through a fine upland meadow, where the walking was easy and the surroundings delightful. By noon we reached a small, shallow lake near the highest part of the divide, considerably below tree line. Here we decided to rest and have lunch. Mr. B. had explored this region with one of his men a few days previously, and from him we learned that we should have to struggle with burnt timber in a few moments. The onward rush of the devastating fire had been stopped near the pass, where the trees were small and scattered. After a short descent we entered the burnt timber. I have never before seen a region so absolutely devastated by fire as this. The fire must have burnt with an unusually fierce heat, for it had consumed the smaller trees entirely, or warped them over till they had formed half circles, with their tops touching the ground. The outcrops of sandstone and quartz rocks had been splintered into sharp-edged, gritty stones, covering the ground everywhere like so many knives. The course of the valley now turned rapidly to the south, so that we rounded a corner of the great mass of mountains culminating in Mount Assiniboine. The mountain itself had been for a long time shut out from view by an intervening lofty ridge of glacier-clad peaks, which were, in reality, merely outlying spurs.

The valley in which we were now walking had an unusual formation, for after a short distance we approached a great step, or drop, whereby the valley bottom made a descent of 400 or 500 feet at an exceedingly steep pitch. Here it was difficult to descend even in the easiest places. Arrived at the bottom of the descent it was not very long before another appeared, far deeper than the first. The mountains on either side, especially a most striking and prominent peak on the east side of the valley, which had hitherto appeared of majestic height, seemed to rise to immeasurable altitudes as we plunged deeper and deeper in rapid descent.

The burnt timber continued without interruption. Our passage became mere log walking, as the extra exertion of jumping over the trees was worse than following a crooked course on top of the prostrate trunks. This laborious and exceedingly tiresome work continued for three hours, and at length the charred trunks, uprooted or burnt off near the ground, and crossed in every direction, were piled so high that we were often ten or twelve feet above the ground, and had to work out our puzzling passage with considerable forethought. At five o'clock our labors ended. We made a camp near a large stream which appeared to take its source near Mount Assiniboine. The only good thing about this camp was the abundance of firewood, which was well seasoned, required but little chopping, and was already half converted into charcoal. Under the shelter of an overhanging limestone ledge we made three lean-tos by supporting our blankets on upright stakes. Black as coal-heavers from our long walk in the burnt timber, seeking a refuge in the rocky ledges of the mountains, and clad in uncouth garments torn and discolored, we must have resembled the aboriginal savages of this wild region. Some thick masses of sphagnum moss, long since dried up, gave us a soft covering, to place on the rough, rocky ground. Our supper consisted of bacon, hard tack, and tea. Large flat stones

A Mountain Bivouac.

laid on a gentle charcoal fire served to broil our bacon most excellently, though the heat soon cracked the stones in pieces.

At eight o'clock we retired to the protection of our shelter. Overhead the starless sky was cloudy and threatened rain. The aneroid, which was falling, indicated that our altitude was only 4,700 feet above the sea. We arose early in the morning; our breakfast was over and everybody ready to proceed at seven o'clock. We were now on the Pacific slope, and, according to our calculations, on one of the tributaries to the north fork of the Cross River, which, in turn, is a tributary to the Kootanie.

We had a plan to explore up the valley from which our stream issued, but beyond that, all was indefinite. It was possible that this valley led around Mount Assiniboine so that we could reach camp in two days. We were, however, certain of nothing as to the geography of the region which we were now entering.

The clouds covered the entire sky and obscured the highest mountain peaks. Worse still, they steadily descended lower and lower, a sign of bad weather. We had, however, but this day in which to see the south side of Mount Assiniboine, and consequently were resolved to do our best, though the chances were much against us. For three hours we followed the stream through the burnt timber, then the country became more open and our progress, accordingly, more rapid. A little after ten o'clock we sat down by the bank of the stream to rest for a few moments, and eat a lunch of hard tack and cold bacon. Such fare may seem far from appetizing to those of sedentary habits, but our tramp of three hours over the fallen trees was equivalent to fully five or six hours walking on a good country road, and what with the fresh mountain air and a light breakfast early in the morning, our simple lunch was most acceptable.

A most pleasing and encouraging change of weather now took place. A sudden gleam of sunlight, partially paled by a thin cloud, called our attention upward, when to our great relief several areas of blue sky appeared, the clouds were rising and breaking up, and there was every prospect of a change for the better.

Once more assuming our various packs, we pushed on with renewed energy. On the left or south was a long lofty ridge of nearly uniform height. On the right was a stupendous mountain wall of great height, the top of which was concealed by the clouds. This impassable barrier seemed to curve around at the head of the valley, and, turning to the south, join the ridge on the opposite side. This then was a "blind" valley without an outlet. There were two courses open to us. The first was to wait a few hours, hoping to see Mount Assiniboine and return to camp the way we came. The second was to force a passage, if possible, over the mountain ridge to the south and so descend into the North Fork valley, which we were certain lay on the other side. The latter plan was much preferable, as we would have a better chance to see Mount Assiniboine, and the possibility of returning to camp by a new route.

After a short discussion, we selected a favorable slope

and began to ascend the mountain ridge. A vast assemblage of obstacles behind us in the shape of two high passes, dense forests, and a horrid infinity of fallen trees, crossed bewilderingly, made a picture in our minds, constant and vivid as it was, that urged us forward. In striking contrast to this picture, hope had built a pleasing air castle before us. We were now climbing to its outworks, and should we succeed in capturing the place, a new and pleasant route would lead us back to camp and place us there—so bold is hope—perhaps by nightfall.

Thus with a repelling force pushing from behind and an attractive force drawing us forward, we were resolved to overcome all but the insuperable.

There was much of interest on the mountain slope, which was gentle, and allowed us to pay some attention to our surroundings. On this slope the scattered pine trees had escaped the fire and offered a pleasant contrast to the burnt timber. We passed several red-colored ledges containing rich deposits of iron ore, while crystals of calcite and siderite were strewed everywhere, and often formed a brilliant surface of sparkling, sharp-edged rhombs over the dull gray limestone. Among the limestones and shales we found fossil shells and several species of trilobites.

In an hour we had come apparently to the top of our ridge, though of course we hardly dared hope it was the true summit. As, one by one, we reached a commanding spot, a blank, silent gaze stole over the face of each. To our dismay, a vertical wall of rock, without any opening whatever, stood before us and rose a half thousand feet higher. Thus were all our hopes dashed to the ground suddenly, and we turned perforce, in imagination, to our weary return over the many miles of dead and prostrate tree trunks that intervened between us and our camp.

The main object of our long journey was, however, at this time attained, for the clouds lifted and revealed the south side of Mount Assiniboine, a sight that probably no other white men have ever seen. I took my camera and descended on a rocky ridge for some distance in order to get a photograph. Returning to where my friends were resting, I felt the first sensation of dizziness and weakness, resulting from unusual physical exertion and a meagre diet. I joined the others in another repast of raisins and hard tack, taken from our rapidly diminishing store of provisions.

Some more propitious divinity must have been guiding our affairs at this time, for while we were despondent at our defeat, and engaged in discussing the most extravagant routes up an inaccessible cliff, our eyes fell on a well defined goat trail leading along the mountain side on our left. It offered a chance and we accepted it. Peyto set off ahead of us while we were packing up our burdens, and soon appeared like a small black spot on the steep mountain side. Having already passed several places that appeared very dangerous, what was our surprise to see him now begin to move slowly up a slope of snow that appeared nearly vertical. We stood still from amazement, and argued that if he could go up such a place as that, he could go anywhere, and that where he went we could follow. We rushed after him, and found the goat trail nearly a foot wide, and the dangerous places not so bad as they seemed. The snow ascent was remarkably steep, but safe enough, and, after reaching the top, the goat trail led us on, like a faithful guide pointing out a safe route. We could only see a short distance ahead by reason of the great ridges and gullies that we crossed. Below us was a steep slope, rough with projecting crags, while, as we passed along, showers of loose stones rolled down the mountain side and made an infernal clatter, ever reminding us not to slip. At one o'clock we stood on the top of the ridge 9000 feet above sea-level, having ascended 4300 feet from our last camp.

The valley of the north fork of the Cross River lay far below, with green timber once more in sight, inviting us to descend. After five minutes delay, for another photograph, we started our descent, very rapidly, at first, in order to get warm. We descended a steep slope of loose debris, then through a long gully, rather rough, and rendered dangerous by loose stones, till at length we reached the grassy slopes, then bushes, finally trees and forests, with a warm summery atmosphere. Here, beautiful asters and castilleias, and beds of the fragrant Linneas, delicate, low herbs with pale, twin flowers, each pair pendent on a single stem, gave a new appearance to the vegetation. In still greater contrast to the dark coniferous forests of the mountain, there were many white birch trees, and a few small maples, the first I have ever seen in the Rockies. In a meadow by the river we feasted on wild strawberries, which were now in their prime.

Near the river we discovered a trail, the first we had seen so far on our journey around Assiniboine. After an hour of walking we came to a number of horses, and soon saw on the other side of the river a camp of another party of gentlemen, who were exploring this region, and had been out from Banff twenty-four days. We forded the river, and found it a little over our knees, but very swift.

A very pleasant half hour was spent at this place, enjoying their hospitality, and then we pushed on. We were now going westward up the valley, which held a straight course of about six miles, and then turned around to the north. The trail being good, we walked very rapidly till nightfall in a supreme effort to reach our camp that night. Having now been on our feet almost continuously for the past fifteen hours, we had become so fatigued that a very slight obstruction was sufficient to cause a fall, and every few minutes some one of the party would go headlong among the burnt timber. We had barely enough provisions for another meal, however, and so we desired to get as near headquarters as possible. At length, nightfall having rendered farther progress impossible, we found a fairly level place among the prostrate trees, and, after a meal of bacon and hard tack, lay down on the ground around a large fire. The night was mild, and extreme weariness gave us sound sleep. After four hours of sleep, we were again on foot at four o'clock in the morning. We marched into camp at 6:30, where the cooks were just building the morning fires, and commencing to prepare breakfast.

Topography.

We were without doubt the first to accomplish the circuit of Mount Assiniboine. By pedometer, the distance was fifty-one miles, which we accomplished in forty-six hours, or less than two days.

Mount Assiniboine is the culminating point of a nearly square system of mountains covering about thirty-five square miles. According to my estimates from angles taken by surveying instruments made on the spot, the mountain is 11,680 feet in height. Later on, however, I learned from Mr. McArthur, who is connected with the Topographical Survey, and who has probably climbed more peaks of the Canadian Rockies than any other two men, that, according to some angles taken on this mountain from a great distance, the height is 11,830 feet.

Three rivers, the Spray, the Simpson, and the North Fork of the Cross, drain this region, and as the two latter flow into the Columbia, and the former into the Saskatchewan, this great mountain is on the watershed, and consequently on the boundary line between Alberta and British Columbia. About two-thirds of the forest area round its base has been burned over, and this renders the scenery most unattractive. The north and northwest sides, however, are covered with green timber, and studded with lakes, of which one is two miles or more in length. There are in all thirteen lakes around the immediate base of the mountain, and some are exquisitely beautiful.

The great height and striking appearance of Mount Assiniboine will undoubtedly, in the future, attract mountaineers to this region, especially as a much shorter route exists than the one we followed. If the trail is opened along the Spray River, the explorer should be able to reach the mountain, with horses, in two days from Banff. Mount Assiniboine, especially when seen from the north, resembles the Matterhorn in a striking manner. Its top is often shrouded in clouds, and when the wind is westerly, frequently displays a long cloud banner trailing out from its eastern side. The mountain is one that will prove exceedingly difficult to the climber. On every side the slope is no less than fifty degrees, and on the east, approaches sixty-five or seventy. Moreover, the horizontal strata have weathered away in such a manner as to form vertical ledges, which completely girdle the mountain, and, from below, appear to offer a hopeless problem. In every storm the mountain is covered with new snow, even in summer, and this comes rushing down in frequent avalanches, thus adding a new source of danger and perplexity to the mountaineer.

The day of our arrival in camp was spent in much-needed rest. Our time was now up, and it was necessary, on the next day, to commence our homeward journey, and, as our winding cavalcade left the beautiful site of our camp under the towering walls of Mount Assiniboine, many were the unexpressed feelings of regret, for in the two weeks spent here we had had many delightful experiences, and had become familiar with every charming view of lakes and forests and mountains.

In two days we reached the fork where the Simpson

and Vermilion rivers unite. It was our intention to follow up the Vermilion River and reach the Bow valley by the Vermilion Pass. The Vermilion River is at this point a large, deep stream flowing swiftly and smoothly The valley is very wide and densely forested, with occasional open places near the river. For three days we progressed up the river, often being compelled to cross it on account of the dense timber. At one place, after several of the horses had gained a bar in the middle of the river, one of those following, got beyond his depth and was swept rapidly down, and appeared in great danger of being drowned. Fortunately, the animal was caught by an eddy current, and by desperate swimming at length gained the bar. The poor beast was, however, so much benumbed by the cold water that he could not climb upon the bar, but the men dashed in bravely, and by pulling on head and packs, and even his tail, the animal finally struggled into shallow water. Standing up to our knees in the water, with a deep channel on either side of us and an angry rapid below, our prospects were far from encouraging.

I mounted old Chiniquy behind Peyto and we plunged in first. "It's swim sure this time," said Peyto to me, as the water rose at once nearly to the horse's back, and the ice-cold water, creeping momentarily higher, gave us a most uncomfortable sensation. The current was so swift that the water was banked up much higher on the upstream side. Such crossings are very exciting, for at any moment the horse may stumble on the rough bottom or plunge into a deep hole. Chiniquy had a hard time of it and groaned at every step, but got us across all right. The rest all followed, not, however, over-confident at our success, to judge by their anxious looks. All got across except one pack-horse, which, after a voyage down stream, we finally caught and pulled ashore.

There was evidence of much game in this valley, as we saw many tracks of deer, caribou, and bears. One day, just as we stopped to camp, a doe started up and ran by us. We camped on August 2nd at a beautiful spot near the summit of the Vermilion Pass. A large stream came in from the northwest, and we set out to explore it for a short distance, as, before leaving Banff, we had heard of a remarkable canyon near this place.

Not more than an eighth of a mile from the junction of the two streams the canyon commences. At first, the stream is hemmed in by two rocky walls a few feet in height, but as one ascends, the walls become higher and higher, and the sound of the roaring stream is lost in the black depth of a gloomy chasm. To one leaning over the edge of the beetling precipice, this wonderful gorge appears like a bottomless rift or rent in the mountain side, and so deep is it and so closely do the opposite, irregular walls press one towards another, that it is impossible to see the waters below from which a faint, sullen murmur comes up.

Most wonderful of all, the canyon at length comes to a sudden termination, and here the whole mighty stream plunges headlong, as it were, into the very bowels of the

A Wonderful Canyon.

earth. The boiling stream, turned snow-white by a short preliminary leap, makes a final plunge downwards and is lost to sight in a dark cavernous hole, perhaps 300 feet deep, whence proceeds a most awful roar, like that of ponderous machinery in motion. The ground, which is here a solid quartzite formation, fairly trembles at the terrible concussion and force of the falling waters, while cold, mist-laden airs ascend in whirling gusts from the awful depths. Niagara is majestically and supremely grand, but this lesser fall, where the water plunges into a black bottomless hole, is by far the more terrifying.

On the fourth of August we reached the summit of the Vermilion Pass. On the summit we passed several small lakes in the forest. The water was of a most beautiful color, far more vivid than any I have hitherto seen. In the shallow places where the bottom could be easily seen, the water assumed a bright, clear, green color, and in the deeper places, according to the light and angle of view, the color

LAKE ON VERMILION PASS.

varied to darker hues of all possible shades and tints. The rich colors of sky and water in the Rocky Mountains is one of the most beautiful features of the scenery, but likewise one that can only be appreciated by actual experience.

Our horses were plagued by great numbers of bull-dog flies as we entered the Bow valley. It seems as though these insects were more numerous in the valley of the Bow, and its various tributaries, than in those parts of the mountains drained by other rivers.

At four o'clock we reached the Bow River, and forded it where the width was about one hundred yards, and the depth four feet. My camera and several plates were flooded in this passage, which was, however, effected in safety.

A march of one hour more, along the tote-road, brought us to the station of Castle Mountain, once a thriving village in the railroad-construction days, but now presenting a forlorn and deserted appearance. The section men flagged the east-bound train for us, and we arrived in Banff that evening, after having been in camp for twenty-nine days.

CHAPTER XI.

The Waputehk Range—Height of the Mountains—Vast Snow Fields and Glaciers—Journey up the Bow—Home of a Prospector—Causes and Frequency of Forest Fires—A Visit to the Lower Bow Lake—Muskegs— A Mountain Flooded with Ice—Delightful Scenes at the Upper Bow Lake—Beauty of the Shores—Lake Trout—The Great Bow Glacier.

THE Summit Range of the Rocky Mountains as they extend northward from the deep and narrow valley of the Kicking Horse River has a special name—the Waputehk Range,—derived, it is said, from a word which in the language of the Stoney Indians means the White Goat.

From the summit of one of the peaks in this range, the climber beholds a sea of mountains running in long, nearly parallel, ridges, sometimes uniting and rising to a higher altitude, and again dividing, so as to form countless spurs and a complicated topography. In this range each ridge usually presents a lofty escarpment and bare precipitous walls of rock on its eastern face, while the opposite slope is more gentle. Here the Cambrian sandstones and shales and the limestones of later ages may be seen in clearly marked strata tilted up, generally, toward the east, though many of the mountains reveal contortions and faults throughout their structure, which indicate the wellnigh inconceivable forces that have here been at work.

The Waputehk Mountains have remained to this day but very little known, and almost totally unexplored, in their interior portions. No passes are known through this range between the Kicking Horse Pass on the south and the Howse Pass on the north. Then another long interval northwards to the Athabasca Pass is said by the Indians to offer an impassable barrier to men and horses. The continuity of the range is well indicated by the fact that, for a distance of one hundred miles, these mountains present only one pass across the range available for horses. The several ridges which form this range rise to a very uniform altitude of 10,000 or 11,000 feet. On Palliser's map of this region, one peak north of the Howse Pass, Mount Forbes, is accredited with an altitude of 13,400 feet, and the standard atlases have for many years placed the altitude of Mount Brown at 16,000 feet, and that of Mount Hooker at 15,700 feet, but there is much reason to doubt that any mountains attain such heights in this part of the Rocky Mountains.

A heavy snowfall, due to the precipitation brought about by this lofty and continuous range, as the westerly winds ascend and pass over it, and the existence of many elevated plateaus, or large areas having gentle slopes, have conspired to make vast *névé* regions and boundless snowfields among these mountains. From the snowfields, long tongues of ice and large glaciers descend into the valleys, and thus drain away the surplus material from the

The Waputehk Range.

higher altitudes. No other parts of the Rocky Mountains, south of Alaska, have glaciers and snowfields which may compare in size or extent with those of the Waputehk Range.

The desolate though grand extent of gray cliffs and boundless snowfields, extending farther than the eye can reach, when seen from a high altitude, gives no suggestion of the delightful valleys below, where many beautiful lakes nestle among the green forests, and form picturesque mirrors for the surrounding rugged mountains. On the shores of one of these mountain lakes, in the genial warmth of lower altitudes, where the water is hemmed in, and encroached upon, by the trees and luxuriant vegetation fostered by a moist climate, the explorer beholds each mountain peak as the central point of interest in every view. Each cliff or massive snow-covered mountain then appears an unscalable height reaching upward toward the heavens,—a most inspiring work of nature, raising the eyes and the thoughts above the common level of our earth. When seen from high altitudes, a mountain appears merely as a part of a vast panorama or a single element in a wild, limitless scene of desolate peaks, which raise their bare, bleak summits among the sea of mountains far up into the cold regions of the atmosphere, where they become white with eternal snow, and bound by rigid glaciers.

Having become much interested in reports of the vast dimensions of the glaciers in the Waputehk Mountains, and the beauty of the lakes, especially near the sources of the Bow River and the Little Fork of the Saskatchewan, I started on August 14th, 1895, with the intention of visiting those regions and spending some time there. My outfit

READY TO MARCH.

consisted of five horses, a cook, and a packer. I had engaged Peyto for the latter service, as he had been most efficient on our trip to Mount Assiniboine. We left Laggan a little before noon. Not far from the station, there commenced an old tote-road, which runs northward for many miles toward the source of the Bow River. This tote-road had been hastily built for wagons, previous to the construction of the railroad through the Kicking Horse Pass, for at one time it was thought the line would cross the range by the Howse Pass.

Thus for several miles we enjoyed easy and rapid travelling. The weather was mild and pleasant, and my men seemed pleased at the prospect of another month or so in camp.

In the course of a few miles we came to the house of an old prospector. As this was the farthest outpost of

Home of a Prospector.

civilization, and the old man was reported to be an interesting character, I entered the log-house for a brief visit. The prospector's name was Hunter. I found him at home and was cordially welcomed. Here, in a state of solitude and absolute loneliness, with no lake or stream to entertain, and surrounded by a bristling maze of bleached bare sticks looking like the masts of countless ships in a great harbor, this man had spent several years of his life, and, moreover, was apparently happy. On his table I saw spread about illustrated magazines from the United States and Canada, newspapers, and books. The house was roughly but comfortably finished inside, and furnished with good chairs and tables evidently imported from civilization.

This isolated dwelling and its solitary inhabitant reminded me somewhat of Thoreau at Walden Pond. Like this lover of nature, Hunter enjoys his hermit life, which he varies occasionally by a visit to the village of Laggan. Hunter had the better house of the two men, but Thoreau must have had much more to entertain him, in his garden, and the beautiful lake with its constant change of light and shadow, and the surrounding forests full of well-known plants and trees, where his bird and animal friends lived in undisturbed possession.

No sooner had we taken leave of this interesting home of the old prospector, than the trail plunged into the intricacies of the burnt timber, and our horses were severely tried. Peyto and another man had been at work on this part of the trail for two days, very fortunately for us, as without some clearing we should not have been able to force our way through.

The fire had run through after the tote-road was built, so that the fallen timber now rendered it nearly impassable in many places. The forest fires have been much more frequent since the country has been opened by the whites, but it would be a great mistake to conclude that before the arrival of civilized men the country was clothed by an uninterrupted primeval forest. When we read the accounts of Alexander Mackenzie, and the earliest explorers in the Rocky Mountains, we find burnt timber frequently mentioned.

However, these accounts only cover the last one hundred years, and records of geology must be sought previous to 1793. Dr. Dawson mentions a place near the Bow River where forest trees at least one hundred years old are growing over a bed of charcoal made by an ancient forest fire. Another bank near the Bow River, not far from Banff, reveals seven layers of charcoal, and under each layer the clay is reddened or otherwise changed by the heat. Thus the oldest records carry us back thousands of years. The cause of these ancient fires was probably, in great part, lightning, and possibly the escaping camp fires of an aboriginal race of men.

Forest fires in the Canadian Rockies only prevail at one season of the year—in July, August, and September,—when the severe heat dries up the underbrush and fallen timber. Earlier than this, everything is saturated by the melting snows of winter, while in autumn the sharp frosts and heavy night dews keep the forests damp. According to the condition of the trees, a forest fire will burn sometimes slowly and sometimes with fearful rapidity. When a long period of dry, hot weather has prevailed, the fire, once started, leaps from tree to tree, while the sparks soar high into the air and, dropping farther, kindle a thousand places at once. The furious uprush of heated air causes a strong draught, which fans the fire into a still more intense heat. Sometimes whirlwinds of smoke and heated air are seen above the forest fires, and at other times the great mass of vapor and smoke rises to such a height that condensation ensues, and clouds are formed. In the summer of 1893, a forest fire was raging about five miles east of Laggan. Standing at an altitude of 9000 feet, I had a grand view of the ascending smoke and vapors, which rose in the form of a great mushroom, or at other times more like a pine tree,—in fact, resembling a volcanic eruption. Judging by the height of Mount Temple, the clouds rose about 13,000 feet above the valley, or to an altitude of 18,000 feet above sea-level. It was a cumulus cloud, shining brilliant in the sunlight, but often revealing a coppery cast from the presence of smoke. The ascending vapors gave a striking example of one of the laws of rising air currents. The tendency of an ascending column of air is to break up into a succession of uprushes, separated by brief intervals of repose, and not to rise steadily and constantly. The law was clearly illustrated by this cloud, which, at intervals of about five or six minutes, would nearly disappear and then rapidly form again and rise to an immense height and magnitude.

In the course of a few years after a forest fire has swept along its destructive course, the work of regeneration begins, and a new crop of trees appears. Sometimes the growth is alike all over the burnt region, young trees springing up spontaneously everywhere, and sometimes the surrounding green forests send out skirmishers, and gradually encroach on the burnt areas. Curiously enough, however, a new kind of tree replaces the old almost invariably. Out on the prairie the poplar usually follows the coniferous trees, but in the Rockies, where the poplar can not grow at high altitudes, the pines follow after spruce and balsam, or *vice versa*. The contest of species in nature is so keen that the slightest advantage gained by any, is sufficient to cause its universal establishment. This is probably due to the fact that the soil becomes somewhat exhausted in the particular elements needed by one species of tree, so that when they are removed by an unnatural cause, new kinds have the advantage in the renewed struggle for existence. Thus we have a natural rotation of crops illustrated in the replacement of forest trees.

While we have been considering the causes and effects of forest fires, our horses and men have been struggling with the more material side of the question, and as the imagination leaps lightly over all sorts of obstacles, let us now overtake them as they arrive at a good camping place about eight miles from Laggan. Here the Bow is no longer worthy the name of a river, but is rather a broad, shallow stream, flowing with moderate rapidity. Towards evening Peyto shot a black duck on the river, and I caught a fine string of trout, so that our camp fare was much improved.

The next day we marched for about three hours through light forests and extensive swamps, finally pitching our camp near the first Bow Lake. The fishing was remarkably fine in this part of the river. From a single pool I caught, in less than three minutes, five trout which averaged more than one pound each. We camped in this place for two days in order to have time to explore about the lake. This first Bow Lake is about four miles long, by perhaps one mile wide, and occupies the gently curving basin of a valley which here sweeps into that of the Bow. There is something remarkable in the unusual manner in which the Bow River divides itself into two streams some time before it reaches this lake. The lesser of these two streams continues in a straight course down the valley, while the larger deviates to the west and flows into the lower end of the lake, only to flow out again about a fourth of a mile farther down, at the extreme end of the lake. The island thus formed is intersected everywhere by the ancient courses of the river, which are now marked by crooked and devious channels, in great part filled with clear water, forming pools everywhere. This whole region must have once formed part of a much larger lake, as for several miles down the valley there are extensive swamps, almost perfectly level and underlaid by large deposits of fine clay.

The drier places in these muskegs are covered with a growth of bushes or clumps of trees, gathered together on hummocks slightly elevated above the general level. A rich growth of grass and sedge covers the lower and wetter places, which often assume all the features of a peat bog, with a thick growth of sphagnum mosses, while the ground trembles, for many yards about, under the tread of men and horses.

The next day Peyto and I crossed the river on one of our best horses known as the " Bay," and after turning him back towards the meadow, we started on a tramp around the lake. We followed the west shore for the entire distance. The last half mile was over a talus slope of loose stones, broken down from the overhanging mountain, and now disposed at a very steep angle. There was a barely perceptible shelf or beach about six inches wide, just at the edge of the water, which we gladly took advantage of while it lasted.

The glacial stream entering the lake has built out a curious delta, not fan-shaped as we should expect, but almost perfectly straight from shore to shore. This delta is a great gravel wash, nearly level, and quite bare of trees or plants, except a few herbs, the seeds of which have lately been washed down from higher up the valley. All this material has been carried into the lake since the time when, in the great Ice Age, these valleys were flooded with glaciers several thousand feet in depth.

As we turned the corner near the end of the gravel wash, the glaciers at the head of the valley began to

appear, and in a few more steps we commanded a magnificent view of a great mountain, literally covered by a vast sheet of ice and snow, from the very summit down to our level. As we looked up the long gentle slope of this mountain, we could hardly realize that it rose more than 5000 feet above us. The glacier which descended into the valley was not very wide, but showed the lines of flow very clearly. Six converging streams of ice united to form the glacier on our right, while the glacier on the left poured down a steep descent from the east, and formed a beautiful ice cascade, where the sharp-pointed *seracs*, leaning forward, resembled a cataract suddenly frozen and rendered motionless. As if by way of contrast, a beautiful little waterfall poured gracefully over a dark precipice of rock on the opposite side of the valley, and added motion to this grand expanse of dazzling white snow. The loud-roaring, muddy stream near where we stood, is one of the principal sources of the Bow, and, after depositing its milky sediment in the lake, the waters flow out purified and crystal clear, of that deep blue color characteristic of glacial water. On a smaller scale this lake is like Lake Geneva, with the Rhone entering at one end, muddy and polluted with glacial clays, and flowing out at the other, transparently clear, and blue as the skies above it.

After a partial ascent of Mount Hector on the next day, we moved our camp and continued our progress up the Bow River for about two hours. Here we camped on a terrace near the water, surrounded on all sides by a very light forest in a charming spot. On the following day the trail led us for two miles through some very bad country, where the horses broke through the loose ground between the roots of trees, and in their efforts to extricate themselves were often in great danger of breaking a leg. Fortunately, however, this was not of long duration. The trail soon improved and became very clearly marked like a well made bridle-path. It led us along the banks of the Bow, through groves of black pine, with a few spruces intermingled. The ascent was constant, though gradual, and our altitude was made apparent by the manner in which the trees grew in clumps, and by the fact that the forests were no longer densely luxuriant, but quite open, so that the horses could go easily among the trees in any direction.

In about three hours after leaving camp, our horses entered an open meadow where the trail deserted us, but there was not the slightest difficulty in making good progress. To the south, a great wall of rock rose to an immense height, one of the lower escarpments of the Waputehk Range, and as we progressed through the pleasant moors a remarkable glacier was gradually revealed, clinging to the cliffs in a three-pronged mass. As, one by one, these branches of the glaciers were disclosed, they appeared first in profile, and owing to the very steep pitch down which the ice was forced to descend, the glacier was rent and splintered into deep crevasses, with sharp pinnacles of ice between, which appeared to lean out over the steep descent and threaten to fall at any moment.

The Upper Bow Lake.

The absence of trees to the north of us, and the general depression of the country in that direction, gave us every indication that we were approaching the Upper Bow Lake, nor were our surmises incorrect, for in a few minutes more of progress, after seeing the glacier, glimpses of water surface were to be had in the near distance among the trees. I went ahead of our column of horses and selected a beautiful site for our camp, on the shore of the lake, only a few yards from the water. The surrounding region was certainly the most charming I have seen in the Rocky Mountains. The lake on which we camped was nearly cut off from the main body of water to the north, by a contraction of the shores to a narrow channel. In fact, it might be regarded as a land-locked harbor of the Upper Bow Lake. Just below our camping place the waters were contracted again, and descended in a shallow rapid to another lake, resting against the mountain side on the south. This latter lake is about three or four feet lower than the others, and appeared to be about two-thirds of a mile in length.

This region, for the artist with pencil and brush, would be a fairy-land of inexhaustible treasures. The shores along these various lakes were of a most irregular nature, and in sweeping curves or sudden turns, formed innumerable coves and bays, no less pleasing by reason of their small extent. Long, low stretches of land, adorned with forest trees, stretched straight and narrow far out into the two larger lakes, their ends dissolving into chains of wooded islands, separated from the mainland by shallow channels of the clearest water. In every direction were charming vistas of wooded isles and bushy shores, while in the distance were the irregular outlines of the mountains, their images often reflected in the surface of the water. The very nature of the shores themselves, besides their irregular contours, varied from place to place in a remarkable manner. In one locality the waters became suddenly deep, the abrupt shores were rocky, and formed low cliffs; in other places the bottom shelved off more gradually, and there would be a narrow beach of sand and small pebbles, ofttimes strewed with the wreckage of some storm,—a massive tree trunk washed upon the beach, or stranded in shallow water near the shore.

There were, moreover, many shallow areas and swampy tracts where a rich, rank growth of water grasses and sedges extended into the lake. Such border regions between the land and water were perhaps the most beautiful and attractive of all the many variations of these delightful shores. The coarse, saw-edged leaves of the sedges, harsh to the touch, are pliant in the gentlest breath of wind. The waving meadows of green banners, or ribbons, rising above the water, uniform in height, and sensitive to the slightest air motion, rustle continuously as the breezes sweep over them, and rub their rough surfaces together.

From this region, wherein were combined so many charming views of nature, with mountain scenery, lakes, islands, and forests, all of the most attractive kind, it proved impossible to move our camp for several days.

During the time that we remained here, our explorations and wanderings took us along all the shores and islands, and up the neighboring mountain slopes. On one of the islands opposite our camp we discovered a small pool of singular formation. The pool was nearly circular, and about ten yards in diameter. The bottom was funnel-shaped, and in the very centre was a black circle—in fact a bottomless hole—apparently connected by dark subterranean channels with the depths of the adjacent lake. Its borders were low and swampy, where the spongy ground quaked as we moved about, and trembled so much that we feared at any moment to be swallowed up. In fact the whole pool became rippled by the movements of its banks.

The glacier opposite was the object of another trip, and this, too, proved interesting. The *névé* on the flat plateau above discharges its surplus ice for the most part by hanging glaciers, which from time to time break off and fall down the precipice. We were often startled both day and night by the thunder of these avalanches. Two tongues of ice, however, effect a descent of the precipice where the slope is less steep, and though much crevassed and splintered by the rapid motion, they reach the bottom intact. Here the two streams, together with the accumulations of ice constantly falling down from above, become welded into a single glacier, which terminates only a short distance from the lake. The most unusual circumstance about this glacier is the fact that the ice is much higher at the very end than a little farther back, so that a great, swelling mound of ice, about 200 feet thick, forms the termination.

About one fourth of a mile below the end of the glacier, on an old moraine ridge now covered over with luxuriant forest, we saw a towering cliff of rock rising above the trees. This proved, on a closer examination, to be a separate boulder, which must have been carried there by the ice a long time ago. It was of colossal proportions, at least sixty feet high, and nearly as large in its other dimensions. From the top we had an extensive view of the lakes and valleys; while at its base we found on one side an overhanging roof, making so complete a shelter, that it was not difficult to imagine this place to have been used by savages, in some past age, as a cave dwelling.

Many years ago, not less than one hundred, the forests on the slopes to the east of the valley had been devastated by a fire. The long lapse of time intervening had, however, nearly obliterated the dreary effects of this destruction. The trees had replaced themselves scatteringly among the dead timber, and attained a large size. The fallen trunks showed the great length of time they had lain on the ground by the spongy, decomposed condition of the wood. Many of the trunks had dissolved into red humus, the last stage of slowly decomposing wood, and the fragments were disposed in lines, bare of vegetation, indicating where each tree had found its final resting-place.

The swampy shores and large extent of water surface

in this region fostered many varieties of gnats, mosquitoes, and other insects, though, fortunately, not in such great numbers as to be very troublesome. In fact, the season of the year was approaching that period when the mosquitoes suddenly and regularly disappear, for some unexplained reason. I have always noticed that in the Canadian Rockies the mosquitoes become much reduced in numbers between the 15th and 20th of August, and after that time cause little or no trouble. In order, however, that there may be no lack of insect pests, nature has substituted several species of small flies and midgets, which appear about this time and follow in a rotation of species, till the sharp frosts of October put an end to all active insect life. Some of these small pests are no less troublesome than the mosquitoes which have preceded them, though they afford a variation in their manner of annoyance, and are accordingly the more endurable.

Along the reedy shores of the lake and sometimes over its placid surface, when the air was quiet toward evening, we often saw clouds of gnats hovering motionless in one spot, or at times moving restlessly from place to place, like some lightless will-o'-the-wisp, composed of a myriad of black points, darting and circling one about another. Nature seems to love circular motion: for just as the stars composing the cloudy nebulæ revolve about their centres of gravity in infinite numbers, moving forever, through an infinity of space; so do these ephemeral creations of our world pass their brief lives in a ceaseless vortex of complicated circles.

On one occasion we built a raft to ferry us across the narrow part of the lake so that we might try the fishing on the farther side. The raft was hastily constructed, and, after we had reached deep water, it proved to be in a state of stable equilibrium only when the upper surface was a yard under water. After a thorough wetting we finally reached the shore, and proceeded to build a more trustworthy craft.

On the 21st of August we moved our camp down to the north end of the lake. Here the nature of the scenery is entirely changed. Whereas the lower end of the lake abounds in land-locked channels and wooded islands, so combined as to make the most pleasing and artistic pictures from every shore, the other part of this lake presents regular shore lines, and everything is formed on a more extensive scale. The north side of the lake is curved in a great arc, so symmetrical in appearance that it seems mathematically perfect, and the eye sweeps along several miles of shore at a single glance as though this were some bay on the sea-coast.

As we neared the north end of the lake, a valley was disclosed toward the west, and an immense glacier appeared descending from the crest of the Waputehk Range. Even at a distance of three or four miles, this glacier revealed its great size. The lower part descended in several regular falls to nearly the level of the lake. In the lower part, the glacier is less than a mile in width, but above, the ice stream expands to three or four miles, and extends back indefinitely, probably ten miles or more.

This Great Bow Glacier had the same position relatively to the lake, as the glacier we visited at the Lower Bow Lake held to that body of water.

A better knowledge of these lakes revealed a striking similarity between them. Each lake occupies a curving valley, which in each case enters the Bow valley from the south. The two lakes are about the same size and nearly the same shape, a long gentle curve about five times longer than broad. At the head of each, though at slightly different distances, are large glaciers. The glacial streams have likewise formed flat gravel washes, or deltas, which have encroached regularly on the lake and formed a straight line from shore to shore, perfectly similar one to another. A further resemblance might be observed in the presence of two talus slopes from the mountain sides, in each case on the south side of the lake, near the delta. The Lower Bow Lake is about 5500 feet above sea-level, while the upper lake is a little more than 6000 feet. The increased altitude has the effect of making the forest more open, and the country more generally accessible, in the region of the upper lake. From one point on the shores of the upper lake, five large glaciers may be counted, the least of which is two miles long, and the greatest has an unknown extent, but is certainly ten miles in length.

Our camp was pleasantly located in the woods not far from the water. After Peyto had put up the tent and got the camp in order, with the horses enjoying a fine pasture, he set off to explore the lake shore toward the

Great Glacier. He returned to camp about five o'clock carrying a fine lake trout which he had caught. This fish was taken near the shore, and was probably a small one compared with those which live in deeper water; nevertheless, it measured twenty-three inches in length, and weighed about seven pounds. The Bow lakes have a reputation for abounding in fish of a very large size. So far as I am aware, no boat has ever sailed these waters, and there is no certainty what size the fish may reach in the deeper parts of the lake. Judging by trout which have been caught in Lake Minnewanka, near Banff, it is very probable that they run as high as thirty or forty pounds.

CAMP AT UPPER BOW LAKE.

The next day, Peyto and I took a lunch with us and spent the entire day exploring and photographing the glacier and its immediate neighborhood. The ice is not hemmed in by any terminal moraine, but shelves down gradually to a thin edge. In fact the termination of the glacier resembles somewhat the hoof of a horse, or rather that of a rhinoceros, the divided portions being formed by crevasses, while long thin projections of ice spread out between. It is a very easy matter to get on the glacier, and quite safe to proceed a long way on its smooth surface. We had some fine glimpses of crevasses so deep that it was impossible to see the bottom, while the rich blue color of the ice everywhere revealed to us marvels of colored grottoes and hollow-sounding caverns, their sides dripping with the surface waters. There is something peculiarly attractive, perhaps from the danger, pertaining to a deep crevasse in a glacier. One stands near the edge and throws, or pushes, large stones into these caverns, and listens in awe to the hollow echoes from the depths, or the muffled splash as the missile finally reaches a pool of water at the bottom. There is a suggestion of a lingering death, should one make a false step and fall down these horrible crevasses, where, wedged between icy walls far below the surface, one could see the glimmering light of day above, while starvation and cold prolong their agonies. A party of three mountaineers thus lost their lives on Mount Blanc in 1820, and more than forty years later their bodies were found at the foot of the Glacier des Bossons, whither they had been slowly transported, a distance of several miles, by the movement of the ice. The most dangerous crevasses are not those of the so-called "dry glacier," where the bare ice is everywhere visible, but those of the *névé* regions where the crevasses are concealed, or obscured by the overlying snow.

Not far from the foot of the glacier the muddy stream flows through a miniature canyon, with walls near together, cut out of a limestone formation. The water here rushes some quarter of a mile, foaming and angry, as it dashes over many a fall and cascade. Where the canyon is deepest an immense block of limestone about twenty-five feet long has fallen down, and with either end resting on the canyon walls, it affords a natural bridge over the gloomy chasm. As probably no human being had ever crossed this bridge, we felt a slight hesitation in making the attempt, fearing that even a slight jar might be sufficient to dislodge the great mass. It proved, however, quite safe and will undoubtedly remain where it is for many years and afford a safe crossing-place for those who visit this interesting region.

CHAPTER XII.

Sources of the Bow—The Little Fork Pass—Magnificence of the Scenery—Mount Murchison—Camp on the Divide—A High Mountain Ascent—Future of the Bow Lakes—Return down the Bow—Search for a Pass—Remarkable Agility of Pack-Horses—The "Bay" and the "Pinto" —Mountain Solitudes—Mount Hector—Difficult Nature of Johnston Creek—A Blinding Snow-Storm—Forty-Mile Creek—Mount Edith Pass.

A FINE trout stream entered the lake near our camp. This was, in fact, the Bow River. It held a meandering course a short distance before entering the lake, through a level meadow, or rather an open region, thickly grown over with alder bushes and other shrubby plants.

We were delayed at this camp by a period of unsettled weather with occasional storms and strong winds, so that three days were required to finish our explorations. At length, on the 24th of August, we broke camp, and followed the Bow valley northwards towards the source of the river. The valley preserves its wide character to the head of the pass, and is unusual among all the mountain passes for several reasons. The ascent to the summit is very gradual and constant, the valley is wide, and the country is quite open. In about two hours we came to the summit, and, after a long level reach, the slope insensibly changed and the direction of drainage was reversed.

This was a most delightful region. The smooth valley bottom sloped gradually upward toward the mountains on the east and west, and insensibly downward toward the valleys north and south, thus making an extensive region with gentle slopes curving in two directions, which in some way impresses the mind with a sense of quiet grandeur and indefinite liberty. But chiefly this region of the divide is made charming by a most beautiful arrangement of the trees. There are no forests here, nor do the trees grow much in groves or clumps, but each tree stands apart, at a long interval from every other, so that the branches spread out symmetrically in every direction and give perfect forms and beautiful outlines. Between are smooth meadows, quite free of brush, but crowded with flowering plants, herbs, and grasses, so that the general impression is that of a gentleman's park, under the control and care of a landscape gardener, rather than of the undirected efforts of nature.

I shall never forget the first view we had into the valley of the Saskatchewan. Approaching a low ridge at the south side of the valley, suddenly there is revealed a magnificent panorama of glaciers, lakes, and mountains, unparalleled among the Canadian Rockies for its combination of grandeur and extent. To the south, one beholds the end of an immense glacier, at the termination of which there are two great arched caverns in the ice. From out these issue two roaring glacial streams, the source of the Saskatchewan River, or at least of its longest tributary called the Little Fork. Lofty mountains

The Little Fork Pass.

hem in this glacier on either side, only revealing a portion of the vast *névé* which may be seen extending southward for six or seven miles.

To the north and, as it were, at our feet, though in reality a thousand feet below, lay a large and beautiful lake with irregular outlines. This lake reaches several miles down the valley of the Little Fork, which here extends northward so straight and regular, that the view is only limited at the distance of thirty miles by the long range of mountains on its east side. Dr. Hector, who came through this region in the fall of 1858, comments on the magnificent extent and grandeur of this view.

Through a notch in a mass of mountains to the north, there appeared the extreme summit of Mount Murchison, a very sharp and angular rock peak, which the Indians regard as the highest mountain of the Canadian Rockies. According to some rough angles taken by Dr. Hector, this mountain has an altitude of 13,500 feet. In Palliser's Papers a sketch of this mountain, as seen from the summit of the Pipestone Pass, makes the rock peak much more sharp and striking in appearance even than that of Mount Assiniboine, or of Mount Sir Donald in the Selkirks.

We continued our journey over the pass and descended into the valley of the Little Fork for several miles. The trail was very good, though the descent was remarkably steep. We camped by a small narrow lake, in reality merely an expansion of the Little Fork. Behind us was an area of burnt timber, but southward the forests were in their primeval vigor and the mountains rose to impressive heights above. The weather became rather dubious, and during the night there was a fall of rain, followed by colder weather, so that our tent became frozen stiff by morning.

It seemed best to return the next day to the summit of the pass, where everything conspired to make an ideal camping place. Accordingly, the men packed the horses and we located our camp on the crest of the divide, 6350 feet above sea-level. The tent was pitched in a clump of large trees surrounded on all sides by open meadows, where one could wander for long distances without encountering rough ground or underbrush. Near the camp a small stream, and several pools of clear water, were all easily accessible.

The next day I induced Peyto to ascend a mountain with me. He was not used to mountain climbing, and had never been any higher than the ridge that we were compelled to cross when we were walking around Mount Assiniboine, which was less than 9000 feet in altitude. The peak which I had now in view lay just to the northeast from our camp on the pass. It appeared to be between 9000 and 10,000 feet high, and offered no apparent difficulties, on the lower part at least. We left camp at 8:30 A.M. and passed through some groves of spruce and balsam, where we had the good fortune to see several grouse roosting among the branches of the trees. Peyto soon brought them down with his six-shooter, in handling which he always displays remarkable accuracy

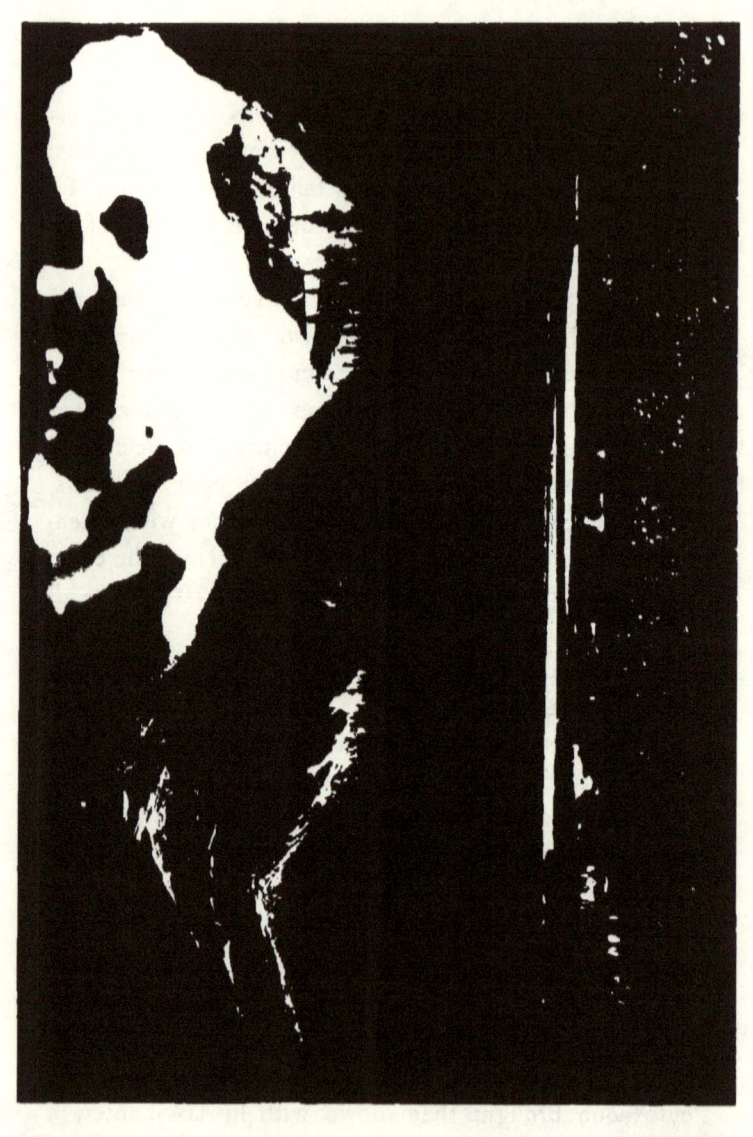

and skill. Many a time, when on the trail, I have seen him suddenly take his six-shooter and fire into a tall tree, whereupon a grouse would come tumbling down, with his neck severed, or his head knocked off by the bullet.

A hawk scented our game and came soaring above us so that we had to hide our birds under a covering of stones, as of course we did not care to take them with us up the mountain. We found not the slightest difficulty in the ascent till we came near the summit. The atmosphere was remarkably clear, and some clouds high above the mountains rendered the conditions very good for photography. At an altitude of 9800 feet we came to the summit of the *arête* which we were climbing, and saw the highest point of the mountain about one-third of a mile distant, and considerably higher. Fortunately, a crest of snow connected the two peaks, and with my ice-axe I knocked away the sharp edge, and made a path. In a few minutes we were across the difficult part and found an easy slope rising gradually to the summit. We reached it at 11:30, and found the altitude 10,125 feet. The view from the great snow dome of this unnamed mountain was truly magnificent. The Waputehk Range could be seen through an extent of more than seventy-five miles, while some of the most distant peaks of the Selkirks must have been more than one hundred miles from where we stood. To the east about ten miles was the high peak of Mount Hector, almost touching the clouds.

In the northern part of the Waputehk Range we saw some very high peaks, though the clouds covered everything above 11,000 feet. There seemed to be a storm in that direction, as snow could be discerned falling on the mountains about thirty miles distant. The general uniformity of height, and the absence of unusually high peaks, a characteristic feature of the Canadian Rockies, were very clearly revealed from this mountain.

Peyto was overwhelmed with the magnificent panorama, and said that he now appreciated, as never before, the mania which impels men to climb mountains. The storm which we saw in the west and north passed over us toward evening, in the form of gentle showers. On the next day, however, the weather was perfectly clear and calm.

On the 26th of August our horses were packed and our little procession was in motion early in the morning, and we were wending our way down the Bow River. I cannot take leave of this region, however, even in imagination, without a word in regard to the unusual attractiveness of this part of the mountains.

In the first place there are magnificent mountains and glaciers to interest the mountaineer, and beautiful water scenes, with endless combinations of natural scenery for the artist; moreover, the streams abound in brook trout and the lakes are full of large lake trout, so numerous as to afford endless sport for fishermen. The botanist, the geologist, and the general lover of science will likewise find extensive fields of inquiry open to him on every side.

The time of travelling required by us to reach the Upper Bow Lake was about nine hours, and this was with heavily laden pack-horses. Hitherto, only those con-

Visions of the Future.

nected with the early explorations, or the railroad surveys, have visited this lake, but I cannot look forward to the future without conjuring up a vision of a far different condition of things. In a few years, if I mistake not, a comfortable building, erected in a tasteful and artistic manner, will stand near the shores of this lake on some beautiful site. A steam launch and row-boats or canoes will convey tourists and fishermen over the broad waters of the lake, and a fine coach road will connect this place with Laggan, so that passengers may leave Banff in the morning and, after a ride of two hours by railroad, they will be transferred to a coach and reach the Upper Bow Lake in time for lunch ! If a good road were constructed this would not be impossible, as the distance from Laggan is only about twenty miles, and the total ascent 1000 feet.

With such visions of the future and the more vivid memory of recent experiences in mind, we took leave of the beautiful sheet of water, and continued on our way down the Bow valley. It was not our purpose, however, to return to Laggan directly, for Wilson had planned an elaborate route, by which some of the wilder parts of the mountains might be visited. This route would lead us over a course of about eighty or one hundred miles through the Slate Mountains and Sawback Range, and eventually bring us to Banff.

We were to follow a certain stream that enters the Bow from the north, but as we were now, and had been for many days, outside the region covered by Dawson's map, it was impossible to feel certain which stream we should take. On our way up the Bow River, Peyto had made exploring excursions into several tributary valleys, but in every case these had proved to be hemmed in by precipitous mountain walls, and guarded at the ends by impassable cliffs or large glaciers.

The second day after leaving the lake we came to a large stream which had not been examined hitherto. Though we were far from certain that this was the stream that had been indicated by Wilson, it seemed best to follow up the valley and see where we should come out. After ascending an exceedingly steep bank, we found easy travelling in a fairly open valley. One fact made us apprehensive that there was no pass out of the valley. There was no sign of a trail on either side of the stream, and none of the trees were blazed. Indian trails exist in almost every valley where an available pass leads over the summit, and where there are no trails the probability is that the valley is blind, or, in other words, leads into an impassable mountain wall. The valley curved around in such a manner that we could not tell what our prospects were, but at about two o'clock we reached a place far above timber line,—a region of open moors, absolutely treeless,—surrounded by bare mountains on every side.

Our tent was pitched in a ravine near a small stream. Immediately after lunch, Peyto and I ascended 1000 feet on a mountain north of the valley with the purpose of discovering a pass. From this point we saw Mount Hector due south, and the remarkable mountain named Mount Molar, nearly due east. Three possible

outlets from the valley appeared from our high elevation. Peyto set off alone to explore a pass toward the north, in the direction of the Pipestone Pass, while I made an examination of a notch toward the east. Each proved impossible for horses, if not for human beings. The third notch lay in the direction of Mount Hector, and together we set out to examine it. A walk of about two miles across the rolling uplands of this high region brought us to the pass. It was very steep, but an old Indian trail proved that the pass was available for horses. The trail appeared more like those made by the mountain goats than by human beings, for it led up to a very rough and forbidding cliff, where loose stones and long disuse had nearly obliterated the path. We spent some time putting the trail in repair, by rolling down tons of loose stones, and making everything as secure as possible.

The next morning was threatening, and gray, watery clouds hung only a little above the summit of the lofty pass, which was nearly 8000 feet above sea-level. I started about an hour before the outfit, as I desired to observe the horses climbing the trail. I felt considerable anxiety as they approached. All my photographic plates, the result of many excursions and mountain ascents in a region where the camera had never before been used, were placed on one of the horses, for which purpose one of the most sure-footed animals had been selected. In case of a false step and a roll down the mountain side, the results of all this labor would be lost.

The horses, however, all reached the summit in safety.

These mountain pack-horses reveal a wonderful agility and sagacity in such difficulties as this place presented. In fact, the several animals in my pack-train had become old friends, for they had been with me all summer. Peyto, as packer, always rode in the saddle, for the dignity of this office never allows a packer to walk, and besides, from their physical elevation on a horse's back they can better discern the trail. A venerable Indian steed, long-legged and lean, but most useful in fording deep streams, was Peyto's saddle-horse. The bell-mare followed next, led by a head-rope. The other horses followed in single file, and never allowed the sound of the bell to get out of hearing. There were two horses in the train that were endowed with an unusual amount of equine intelligence and sagacity. The larger of the two was known as the "Bay," and the other was called "Pinto," the latter being a name given to all horses having irregular white markings. These animals were well proportioned, with thick necks and broad chests, and, though of Indian stock, they probably had some infusion of

THE "BAY."

Spanish blood in their veins, derived from the conquest of Mexico.

The Pinto was remarkably quick in selecting the best routes among fallen timber, or in avoiding hidden dangers, but the Bay was far more affectionate and fond of human company. In camp, all the horses would frequently leave the pasture and visit the tent, where they would stand near the fire to get the benefit of the smoke when the flies were thick, or nose about in the hope of getting some salt. On the trail, it was always very interesting to watch the Bay and Pinto. They would unravel a pathway through burnt timber in a better manner than their human leaders, and would calculate in every case whether it were better to jump over a log or to walk around it. But one day I was surprised to see the Bay jump over a log which measured 3 feet 10 inches above the ground. With a heavy, rigid pack this is more of a feat than to clear a much greater height with a rider in the saddle. Sometimes when the trail was lost we would put the Pinto ahead to lead us, and on several occasions he found the trail for us.

The summit of the pass revealed to us one of those lonely places among the high mountains where silence appears to reign supreme. We were in an upland vale, where the ground was smooth and rolling, and carpeted with a short growth of grass and herbs. On either side were bare cliffs of limestone, unrelieved by vegetation or perpetual snow. Here no birds or insects broke the silence of the mountain solitude, no avalanche thundered among the mountains, and even the air was calm and made no sound in the scanty herbage. All was silent as the desert, or as the ocean in a perfect calm. The dull tramp of our horses, and the tinkling of the bell, were the only sounds that interrupted the death-like quiet of the place. It is said that such places soon drive the lost traveller to insanity, but in company with others these lonely passes afford a delightful contrast to the life and motion and sound of lower altitudes.

As we advanced and commenced to descend, the north side of Mount Hector began to appear. It was completely covered with a great ice sheet and snow fields. Mount Hector is a little more than 11,000 feet in altitude, and gives a good example of how the exposure to the sun affects the size of glaciers in these mountains. On the south and west sides of Mount Hector there is almost no snow, while the opposite slopes are flooded by a broad glacier many miles in area, and brilliant in a covering of perpetual snow.

At the tree line a trail appeared, and led us in rapid descent to the valley. The scenery on all sides was magnificent. Many waterfalls came dashing down from the melting glaciers of Mount Hector and joined a torrent in the valley bottom. The great cliffs about us, and the lofty mountains, visible here and there through avenues in the giant forest trees, were illumined by a brilliant sun, ever now and again breaking through the clouds. About eleven o'clock we stopped to have a light lunch, as was our custom on all long marches. Peyto loosed the girdle of the horses, slipped off the packs, and turned the animals into a meadow near by. Meanwhile our cook cut firewood and made a large pot of tea, which always proved the most acceptable drink when a long march had made us somewhat weary. These brief rests of about forty minutes in the midst of a day's march always proved very beneficial to men and horses.

A long straight valley led us southwards for many miles. In every clear pool or stream, trout could be seen darting about and seeking hiding-places, though we had no time to stop and catch them. At about one o'clock we reached the Pipestone Creek and obtained a view of Mount Temple and other familiar peaks about fifteen miles to the south.

We camped near the stream in a meadow, not far from the Little Pipestone Creek. As the march of this day had brought us back to the region covered by the map, we had little apprehension of losing our way in the future.

The next day we followed up the Little Pipestone Creek and enjoyed a fine trail through a dense forest. We camped near the summit of a pass south of Mount Macoun, which I partially ascended after lunch. The rugged peak named Mount Douglas lay due east, and presented some very large and fine glaciers.

Our camp was on a little peninsula jutting out into a lake, with water of a most brilliant blue color. The sunset colors this evening were heightened by the presence of a little smoke in the atmosphere, which gave a deep copper color to the western sky, while the placid lake appeared vividly blue in the evening light.

The following day, which was the first of September, we continued south over a divide and into the valley of Baker Creek, which we followed for several hours, and then took a branch stream which comes in from the east, and finally camped in a high valley. We were now in the Sawback Range, where the mountains are peculiarly rugged, and the strata thrown up at high angles. The weather was giving evidence of an approaching storm, and before we had made camp the next day in Johnston's Creek, rain began to fall.

Hitherto the nature of the country since leaving the Upper Bow Lake had been such as to render the travelling very easy and delightful, but from this point on, we met with all sorts of difficulties. In the lower part of Johnston's Creek, and in the valley of a tributary which comes in from the northeast, the trail was covered by fallen timber, and our progress was very slow and tedious. Moreover, the weather now became very bad, and we were caught near the summit of a pass between Baker Creek and Forty-Mile Creek in a heavy snow-storm, so that the trail was soon obliterated and the surrounding mountains could not be seen. Fearing that we might lose our bearings altogether, Peyto urged forward the horses at a gallop, so that we might get over the pass before the snow gained much depth.

The descent into the valley of Forty-Mile Creek was very steep, and we camped among some large trees with several inches of snow on the ground. The next day we urged our horses on again and followed down the valley of Forty-Mile Creek. In some parts of the valley we found absolutely the worst travelling I have anywhere met with in the Rockies. The horses were compelled to make long detours among the dead timber, and the axe was frequently required to cut out a passage-way. Frequent snow showers swept through the valley, and, though very beautiful to look at, they kept the underbrush covered with damp snow and saturated our clothes with water.

In the afternoon we reached the summit of the Mount Edith Pass, and once more caught sight of the Bow valley and the flat meadows near Banff. A fine wide trail or bridle-path, smooth and hard, led us down toward the valley. The contrast to our recent trails was very striking. We walked between a broad avenue of trees, each one blazed to such an extent that all the bark had been removed on one side of the tree, and some were practically girdled. This was very different from our recent experience where we had only found a small insignificant axe-mark on some dead tree, about once in every quarter mile, or often none at all during hours of progress.

On the fifth of September we reached Banff late in the evening, and found that the valley was free of new snow by reason of its lower altitude. We had been out for twenty-three days and had covered, in all, about one hundred and seventy-five miles.

CHAPTER XIII.

HISTORICAL.

Origin and Rise of the Fur Trade—The Coureurs des Bois and the Voyageurs—Perils of the Canoe Voyages—The Hudson Bay Company and the Northwest Company—Intense Rivalry—Downfall of the Northwest Company—Sir Alexander Mackenzie—His Character and Physical Endowments — Cook's Explorations — Mackenzie Starts to Penetrate the Rockies—The Peace River—A Marvellous Escape—The Pacific Reached by Land—Perils of the Sea and of the Wilderness.

THE history of the early explorations in the Canadian Rockies centres about the fur trade. From the date of the very earliest settlements in Canada, the quest of furs had occupied a position of chief importance, to which the pursuits of agriculture, grazing, or manufacture had been subordinate. The search for gold, which throughout the history of the world has ever been one of the most powerful incentives to hardy adventure and daring exploit, did not at first occupy the attention of those who were ready to hazard their lives for the sake of possible wealth quickly acquired.

The unremitting and often ruthless destruction of the fur-bearing animals, in the immediate vicinity of the settlements, caused them to become exceedingly scarce, and at length to disappear altogether. But fortunately it was not difficult to induce the Indians to bring their furs from more distant regions, until at length even those who lived in the most remote parts of Canada became accustomed to barter their winter catch at the settlements.

As the trade gradually became more extensive, there sprang up two slightly different classes of men, the *coureurs des bois*, or wood rangers, and the *voyageurs*, each of Canadian birth, but who, by reason of constant contact with the Indians and long-continued separation from the amenities and refinements of civilized life, came at length to have more in common with the rude savages, than with the French settlers from whom they were sprung. Many of these wilderness wanderers married Indian wives, and, moreover, their plastic nature, a result of their French extraction, helped them quickly to assume the manners and customs of the swarthy children of the forest. The *voyageurs*, like the *coureurs des bois*, were accustomed to take long canoe voyages, under the employ of some fur company, or even of private individuals; sometimes alone, but more often several banded together, carrying loads of ammunition, provisions, and tobacco from the settlements and returning with their canoes laden down with beaver, marten, and other furs collected among the Indians. The vast domain of Canada is so completely watered by a network of large streams, rivers, and lakes, more or less connected, that it is not difficult to make canoe voyages in almost any direction throughout the length and breadth of this great territory. It is indeed possible to start from Montreal and journey by water to

Hudson Bay, the Arctic Ocean, or the base of the Rocky Mountains.

The *voyageurs* were a hardy race, possessed of incredible physical strength and untiring patience, remarkable for an implicit obedience to their superiors, and endowed with a happy, careless nature, regardless of the morrow, so long as they were well-off to-day. While making their long and arduous journeys, the *voyageurs* would arouse their flagging spirits with merriment and laughter, or awaken echoes from the wooded shores and rocky cliffs along the rivers and lakes, by their characteristic songs, to the accompaniment of the ceaseless and rhythmic movement of their paddles.

How much of romance and poetry filled up the measure of their simple lives! Nature in all its beauty and grandeur was ever around them, and nature's people—the Indians—were those with whom they most associated. They loved all men, and all men loved them, whether civilized or barbarian. The stranger among them was called Cousin, or Brother, and the great fur barons, the partners in the fur companies, on whom they gazed with awe and admiration, as they travelled in regal state from post to post, and to whom they bore almost the relation of serf to feudal lord, they called by their Christian names. The melodies which they chanted in unison as they glided along quiet rivers, with banks of changing outlines and constant variety of forest beauty, would hardly cease as they dashed madly down some roaring, snow-white rapid, beset with dangerous rocks, where a single false stroke would be fatal. For many days continuously they were wont to travel, with short time for sleep, working hour after hour at the paddle, or making the toilsome portages, when they were accustomed to carry on their backs loads of almost incredible weight. Nevertheless, on any opportunity for relaxation, they were ever ready for revelry, music, and the dance, which they would prolong throughout the night.

The usual dress of the *voyageur* consisted of a coat or capote cut from a blanket, a cotton shirt, moccasins, and leather or cloth trousers, held in place by a belt of colored worsted. A hunting knife and tobacco-pouch, the latter a most indispensable adjunct to the happiness of the *voyageur*, were suspended from his belt. Sometimes they would be absent from the settlements twelve or fifteen months, and many never returned from their perilous trips. Some were drowned while attempting to run dangerous rapids. Others were overtaken by the approach of winter, or were stopped by ice-bound rivers impossible to navigate, and perished miserably from exposure and starvation.

Those who returned, however, would be amply rewarded by the wealth suddenly acquired from the result of their long toil. The dissipation of their gains in the course of a few weeks, accompanied by all manner of revelry, licentiousness, and mad extravagance, was their compensation for long periods of privation. At length, their means being exhausted, a longing for the old manner of life returned, and with renewed hopes they would recommence their long journeys into the wilderness.

The value of the fur trade soon aroused the attention

of a number of wealthy and influential traders, and in 1670 a charter was granted to Prince Rupert and a company of fourteen others, to "the sole trade and commerce" throughout all the regions watered by streams flowing into Hudson or James Bay. This region was henceforth known as Rupert's Land. In addition to the right of trade, the Hudson Bay Company had the authority of government and the dispensation of justice throughout this vast territory.

During the winter of 1783-4, however, a number of Canadian merchants, previously engaged in the fur trade, joined their several interests, and formed a coalition which assumed the name of the Northwest Company.

This organization, governed, as it was, by different principles from that of the Hudson Bay Company, soon became a powerful rival. The younger men in the Northwest Company were fired with ambition and assured of an adequate reward for their services. While for many years their older rivals had slumbered, content with the limits of their territory, the more enterprising Northwest Company, with infinite toil and danger, extended their posts throughout the interior and western parts of Canada, and opened up a new and hitherto undeveloped country. Another great advantage that the Northwest Company had over the Hudson Bay Company resulted from their employment of the suave and plastic *voyageurs*, in whose blood the French quality of ready adaptability to surroundings was especially well shown in their dealings with the Indians, with whom they had the greatest influence.

Rivalry of Hudson Bay Company.

On the other hand, the greater part of the Hudson Bay canoe men were imported from the Orkney Islands. What with their obstinate, unbending nature, and mental sluggishness, these men presented a most unfavorable contrast to the genial *voyageurs*.

The establishment of the Northwest Company aroused the utmost jealousy and animosity of the Hudson Bay Company. While the various parties were engaged in dealings with the Indians, there not infrequently occurred open conflicts, bloodshed, and murder among the agents, in their attempts to outwit and circumvent one another.

At length the partners of the Northwest Company in the interior of Canada, realizing that all the profits were more than balanced by their endless and painful contest, determined to open a negotiation with their rivals, and for this purpose sent two delegates to London with full authority to close whatever agreement would be for the best interests of the company. Just at this time the directors of the two companies were about to sign a contract most favorable to the Northwest Company. Unfortunately, on the eve of this event, the two delegates from Canada made their appearance, and instead of communicating at once with their own directors, they showed their papers to the officers of the Hudson Bay Company. The Hudson Bay Company took advantage of the opportunity, and, instead of receiving terms from the other, now proceeded to dictate them. The outcome of this unfortunate manœuvre was, that the Northwest Company became merged in that of the Hudson Bay Company, together with the privileges and trade of all of the vast territory which the Northwest Company had developed by superior enterprise. Thus, in 1821, the Northwest Company ended its career.

The Hudson Bay Company's territory was at length, from time to time, encroached upon as the colonies of British Columbia, Vancouver's Island, and Manitoba were established. Finally, in 1869, the Company ceded all their governmental and territorial rights to the Dominion, receiving £300,000 in compensation. Their forts or posts, together with a small amount of land in the immediate vicinity, were reserved by them. The Hudson Bay Company still exists as a commercial organization, carrying on a thriving business in many of the principal cities and towns of Canada.

So much by way of introduction to the exploration of the Canadian Rockies.

Let us now turn to Sir Alexander Mackenzie, the hardy explorer who first crossed the continent of North America, after penetrating the grim and inhospitable array of mountains which had hitherto presented an impassable barrier to all further westward progress.

Mackenzie was born in the northern part of Scotland, in the picturesque and historic town of Inverness. The year of his birth is usually set down as 1755. In his youth he emigrated to Canada, and found employment as a clerk to one of the partners in the great Northwest Fur Company. Later on he went to Fort Chipewyan, on Lake Athabasca, and became one of the principal partners in the Northwest Company.

Mackenzie was endowed by nature with a powerful physique and a strong constitution, which enabled him to undergo the unusual hardships of his explorations in the wilderness. Beside these physical qualifications, he was inspired with the ambition necessary to the formation of great plans, and with an enterprising spirit which impelled him to carry them through to a successful termination. Great versatility of idea enabled him to oppose every novel and sudden danger with new plans, while a rugged perseverance, indomitable patience, and a boldness often bordering on recklessness, carried him through all manner of physical and material obstacles. In his dealings with the Indians and his own followers, he showed an unusual tact, a quality which more than any other contributed to his success. Nothing so quickly saps the strength and tries the courage of the explorer, be he ever so bold and persevering, as cowardice and unwillingness among his followers.

Nevertheless, Mackenzie was not a scientific explorer. Outside of the manners and customs of the various tribes with which he came in contact, only the most patent and striking phenomena of the great nature-world impressed him. No better idea of his views on this subject could be obtained than from a passage in the preface to his *Voyages*:

"I could not stop," says Mackenzie, "to dig into the earth, over whose surface I was compelled to pass with rapid steps; nor could I turn aside to collect the plants which nature might have scattered on the way, when my thoughts were anxiously employed in making provision for the day that was passing over me. I had to encounter perils by land and perils by water; to watch the savage who was our guide, or to guard against those of his tribe who might meditate our destruction. I had, also, the passions and fears of others to control and subdue. To-day, I had to assuage the rising discontents, and on the morrow, to cheer the fainting spirits of the people who accompanied me. The toil of our navigation was incessant, and oftentimes extreme; and, in our progress overland, we had no protection from the severity of the elements, and possessed no accommodations or conveniences but such as could be contained in the burden on our shoulders, which aggravated the toils of our march, and added to the wearisomeness of our way.

"Though the events which compose my journals may have little in themselves to strike the imagination of those who love to be astonished, or to gratify the curiosity of such as are enamoured of romantic adventures; nevertheless, when it is considered that I explored those waters which had never before borne any other vessel than the canoe of the savage; and traversed those deserts where an European had never before presented himself to the eye of its swarthy natives; when to these considerations are added the important objects which were pursued, with the dangers that were encountered, and the difficulties that were surmounted to attain them, this work will, I flatter myself, be found to excite an interest and conciliate regard in the minds of those who peruse it."

Thus Mackenzie writes in the preface to his journal. Nevertheless, there is no evidence throughout his works that he was learned or even interested in the sciences of botany or geology. The scientific mind becomes so much absorbed in the search for information, when surrounded by the infinite variety of nature's productions, especially in regions hitherto unknown, that mere inconvenience, physical suffering, or imminent peril is incapable of withdrawing the attention from the chosen objects of pursuit. Whoever reads Humboldt's narrative of travels in the equinoctial regions of South America, especially that part which pertains to his voyage on the Orinoco, will appreciate the truth of this. The stifling, humid heat of a fever-laden atmosphere, the ever present danger of sudden death from venomous serpents, ferocious alligators, or the stealthy jaguar, the very air itself darkened by innumerable swarms of mosquitoes and stinging insects, with changing varieties appearing at every hour of the day and night, were unable to force this great naturalist to resign his work.

Unfortunately, the explorer and the naturalist are not often combined in one person, notwithstanding that the fact of being one, implies a tendency toward becoming the other.

Mackenzie mentions one or two attempts previous to 1792 to cross the Rocky Mountains. No record of these expeditions is available, a circumstance that implies their termination in failure or disaster.

Up to this time the Rocky Mountains, with their awful array of saw-edged peaks covered with a dazzling white mantle of perpetual snow, had stood as the western limit of overland exploration, beyond which no European had ever passed. The Pacific Coast had already been explored by Captain Cook in 1778, and a few years later so accurately charted by Vancouver, that his work is still standard among navigators. The eastern border of the Rockies was vaguely located, but between these narrow strips there remained a vast region, four hundred miles wide, extending to the Arctic Ocean, about which little or nothing was known.

As in the case of other unexplored regions, there were vague and conflicting rumors among the Indians concerning the dangers of these upland fastnesses, accounts of hostile tribes, men partly human, partly animal in form and nature, and colossal beasts, endowed with fabulous strength and agility, from which escape was next to impossible. These Indian tales, though in great part the product of imagination or superstition, unfortunately did but partial justice to the reality, for although the reported dangers and terrors were mythical, there were real and material obstacles in the form of mountain ranges bewildering in their endless extent and complexity, between which were valleys blocked by fallen timber, and torrential streams rendered unnavigable by roaring rapids or gloomy canyons of awful depth. In fact, this region was one of the most difficult to penetrate and explore that the world could offer at that time.

Nevertheless, Mackenzie now turned his attention toward this region, resolved to traverse and explore it till he should reach the Pacific. Moreover, he was confident of success, perhaps realizing his many qualifications for such an enterprise, and certainly encouraged by the remembrance of the difficulties he had overcome during his former voyage, in 1789, to the mouth of that great river which bears his name.

Leaving Fort Chipewyan on Lake Athabasca, he soon reached that great waterway, the Peace River, and with several canoes began to stem the moderate current of this stream, which is at this point about one fourth of a mile in width and quite deep.

The origin of names is always interesting, and that of the Peace River is said to be derived from a circumstance of Indian history. The tribe of Indians called the Knisteneux, who originally inhabited the Atlantic seaboard and the St. Lawrence valley, migrated in a northwesterly direction. In the course of this tribal movement, after reaching the centre of the continent, they at length came in contact with the Beaver Indians, and a neighboring tribe called the Slaves, at a point some fifty leagues due south from Lake Athabasca. The Knisteneux drove these tribes from their lands, the Slave Indians moving northward down the Slave River to Great Slave Lake, from which circumstance the lake derives its name. The term Slave was not applied to indicate servitude, but by way of reproach on their unusual barbarity and destitution. The Beaver Indians moved in another direction, more to the westward, and on the ratification of peace between them and the Knisteneux, the Peace River was assigned as the boundary between them.

After proceeding for three weeks up the Peace River, Mackenzie camped for the winter at a point previously decided on, and early in the following spring recommenced his "voyage," as these inland water journeys are called. Mackenzie was accompanied by Alexander Mackay, one of the officers of the Northwest Company. The crew consisted of six Canadian *voyageurs*, and the party was completed by two Indians, who, it was intended, should act as interpreters and hunters. A single canoe, twenty-five feet long and not quite five feet in extreme breadth, served to carry the entire party, in addition to three thousand pounds of baggage and provisions.

It would be entirely aside from our purpose to narrate in detail the many interesting adventures and narrow escapes of the party. A single incident will serve to throw some light on the perils and toils that were encountered. At the time of the incident in question, they had crossed the watershed by following the south branch of the Peace River to its source, and were now descending a mad torrent which runs westward, and is tributary to the Fraser River, which latter Mackenzie mistook for the Columbia.

It was on the morning of the 13th of June, and the canoe had proceeded but a short distance, when it struck, and, turning sidewise, broke on a stone. Mackenzie and all the men jumped into the water at once, and endeavored to stop the canoe and turn it round. But almost immediately she was swept into deeper water, where it became necessary for everybody to scramble aboard with the greatest celerity. In this uncertain contest, one of the men was left in mid-stream to effect a passage to shore in the best way he could.

"We had hardly regained our situations," writes Mackenzie, "when we drove against a rock, which shattered the stern of the canoe in such a manner that it held only by the gunwales, so that the steersman could no longer keep his place. The violence of this stroke drove us to the opposite side of the river, which is but narrow, when the bow met with the same fate as the stern. At this moment the foreman seized on some branches of a small tree, in the hope of bringing up the canoe, but such was their elasticity that, in a manner not easily described, he was jerked on shore in an instant, and with a degree of violence that threatened his destruction. But we had no time to turn from our own situation to inquire what had befallen him; for, in a few moments, we came across a cascade, which broke several large holes in the bottom of the canoe, and started all the bars, except one behind the scooping seat. If this accident, however, had not happened, the vessel must have been irretrievably overset. The wreck becoming flat on the water, we all jumped out, while the steersman, who had been compelled to abandon his place, and had not recovered from his fright, called out to his companions to save themselves. My peremptory commands superseded the effects of his fear, and they all held fast to the wreck; to which fortunate resolution we owed our safety, as we should otherwise have been dashed against the rocks by the force of the water, or driven over the cascades. In this condition we were forced several hundred yards, and every yard on the verge of destruction; but, at length, we most fortunately arrived in shallow water and a small eddy, where we were enabled to make a stand, from the weight of the canoe resting on the stones, rather than from any exertions of our exhausted strength. For, though our efforts were short, they were pushed to the utmost, as life or death depended on them."

At this juncture, the Indians, instead of making any effort to assist the others, sat down and shed tears, though it is considered a mortal disgrace among Indians to weep except when intoxicated.

On the 22d of July, after encountering countless trials and the dangers of savage foes, no less than the obstacles of nature, Mackenzie reached an arm of the sea in latitude 52° 20′ 48″, where on a rocky cliff he inscribed this brief legend in vermilion: "Alexander Mackenzie from Canada by land, the 22d of July, one thousand seven hundred and ninety-three."

The next day, when alone, he was nearly murdered by a band of Indians, but escaped by his agility and by a fortunate momentary hesitation on the part of the savages.

Mackenzie's return journey was over the same route that he had first taken, and required but four weeks to traverse the mountains.

Perils of the Wilderness.

In reading a detailed account of this voyage, one is impressed with the many perils encountered, no less than the ofttimes remarkable and fortunate escapes from them. It is so with the journals of nearly all great travellers. They recount an endless succession of dangers and adventures by sea and land, from which, though often in the very jaws of death by reason of the operations of nature and the elements, the traveller ever eventually escapes, apparently in defiance of the laws of chance and probability. But we must bear in mind the great host of travellers who have never returned, and whose unfinished journals are lost forever to mankind.

The remotest corners of the earth have been mute witnesses to these tragedies. The inhospitable, rock-bound shores of lonely islands, or low-lying sands of coral reefs, where the ceaseless ocean billows thunder in everlasting surf, have beheld the expiring struggles of many a bold navigator. The colossal bergs and crushing ice of polar seas; hurricanes and typhoons in tropic latitudes; the horrors of fire at sea; the broad wastes of continents; trackless desert sands, where, under a scorching sun, objects on the distant horizon dance in the waving air, and portray mirage pictures of lakes and streams to the thirsty traveller; deep, cool forests bewildering in the endless maze of trees; piercing winter storms, with cutting winds and driving snows; the blood-thirsty pack of famishing wolves; rivers, dangerous to navigate, with impetuous current swirling and roaring in fearful rapids,—all these have their records of death and disaster.

But of them all, man has ever been the worst destroyer. The hostile savage, the mutinous crew, or treacherous guide have proved far more cruel, revengeful, and cunningly destructive than the catastrophes of nature, whose mute, dead forces act out their laws in accordance with the great plan of the universe, unguided by motives of hate, and envy, and the wicked devices of human passions.

CHAPTER XIV.

HISTORICAL.

Captain Cook's Explorations—The American Fur Company—First Exploration of the Fraser River—Expedition of Ross Cox—Cannibalism —Simplicity of a Voyageur—Sir George Simpson's Journey—Discovery of Gold in 1858—The Palliser Expedition—Dr. Hector's Adventures— Milton and Cheadle—Growth of the Dominion—Railroad Surveys— Construction of the Railroad—Historical Periods—Future Popularity of the Canadian Rockies.

THE early explorations of Captain Cook had an almost immediate effect on the development of the fur trade. Upon the publication of that wonderful book, *Cook's Voyages round the World*, wherein were shown the great value and quantity of furs obtainable along the northwest coast of America, a considerable number of ships were fitted out for the purpose of carrying on this trade. Three years after, or in 1792, there were twenty American vessels along the Pacific Coast, from California northward to Alaska, collecting furs, especially that of the sea otter, from the natives.

Of these "canoes, large as islands, and filled with white men," Mackenzie had heard many times from the natives met with on his overland journey across the Rocky Mountains. Mackenzie's journal was not published till 1801.

In this book, however, he outlines a plan to perfect a well regulated trade by means of an overland route, with posts at intervals along the line, and a well established terminus on the Pacific Coast. Should this plan be carried out, he predicted that the Canadians would obtain control of the fur trade of the entire northern part of North America, and that the Americans would be compelled to relinquish their irregular trade.

While the agents of the American Fur Company, a rival organization controlled and managed by Mr. John Jacob Astor, were preparing to extend their limits northwards from their headquarters at the mouth of the Columbia, the Northwest Company was pushing southward through British Columbia, and had already established a colony called New Caledonia near the headquarters of the Fraser River. Thus Mr. Astor's scheme of gaining control of the head waters of the Columbia River was anticipated. The war of 1812 completely frustrated his plans, when the post of Astoria fell temporarily into the hands of the English.

A very good idea of the hardships of life at one of these western posts, together with a brief account of the first exploration of the Fraser River, may be obtained from a letter written in 1809 by Jules Quesnel to a friend in Montreal. The letter is dated New Caledonia, May 1st, 1809, and after a few remarks on other matters, Mr. Quesnel goes on to say: " There are places in the north where, notwithstanding the disadvantages of the country in general, it is possible sometimes to enjoy one's self; but here nothing is to be found but hardship and loneliness. Far away from every one, we do not have the pleasure of getting news from the other places. We live entirely upon salmon dried in the sun by the Indians, who also use the same food, for there are no animals, and we would often be without shoes did we not procure leather from the Peace River.

"I must now tell you that I went exploring this summer with Messrs. Simon Fraser and John Stuart, whom you have met, I believe. We were accompanied by twelve men, and with three canoes went down the river, that until now was thought to be the Columbia. Soon finding the river unnavigable, we left our canoes and continued on foot through awful mountains, which we never could have passed had we not been helped by the Indians, who received us well. After having passed all those bad places, not without much hardship, as you may imagine, we found the river once more navigable, and got into wooden canoes and continued our journey more comfortably as far as the mouth of this river in the Pacific Ocean. Once there, as we prepared to go farther, the Indians of that place, who were numerous, opposed our passage, and we were very fortunate in being able to withdraw without being in the necessity of killing or being killed. We were well received by all the other Indians on our way back, and we all reached our New Caledonia in good health. The mouth of this river is in latitude 49°, nearly 3° north of the real Columbia. This trip procured no advantage to the company, and will never be of any, as the river is not navigable. But our aim in making the trip was attained, so that we cannot blame ourselves in any manner."

This letter throws some light on the history of this period, and shows whence the names of certain rivers and lakes of British Columbia were derived. It would be in place here to say that when Mackenzie first came to the Fraser River, after crossing the watershed from the Peace River, he entertained the idea that he was on the Columbia.

A few years later, the agents of the fur companies had established certain routes and passages across the mountains, which they were accustomed to follow more or less regularly in their annual or semi-annual journeys. One of the largest of these early parties to traverse the Rockies was under the management of Mr. Ross Cox, who was returning from Astoria in the year 1817. There were, in all, eighty-six persons in his party, representing many nationalities outside of the various Indians and some Sandwich Islanders.

A striking incident in connection with this expedition illustrates the hazard and danger which at all times attended these journeys through the wilderness. The party had pursued their way up the Columbia River, and were now on the point of leaving their canoes and proceeding on foot up the course of the Canoe River, a stream that flows southward and enters the Columbia not far from the Athabasca Pass. The indescribable toil of their passage up the Columbia, and the many laborious portages, had sapped the strength of the men and rendered some of them wellnigh helpless. Under these circumstances, it seemed best that some of the weakest should not attempt to pursue their journey farther, but should return down the Columbia. There were seven in this party, of whom only two were able to work, but it was hoped that the favorable current would carry them rapidly towards Spokane, where there was a post established. An air of foreboding and melancholy settled upon some of those who were about to depart, and some prophesied that they would never again see Canada, a prediction that proved only too true. In Ross Cox's *Adventures on the Columbia River* the record of their disastrous return is thus vividly related:

"On leaving the Rocky Mountains, they drove rapidly down the current until they arrived at the Upper Dalles, or narrows, where they were obliged to disembark. A cod-line was made fast to the stern of the canoe, while two men with poles preceded it along the banks to keep it from striking against the rocks. It had not descended more than half the distance, when it was caught in a strong whirlpool, and the line snapped. The canoe for a moment disappeared in the vortex, on emerging from which it was carried by the irresistible force of the current to the opposite side, and dashed to pieces against the rocks. They had not had the prudence to take out either their blankets or a small quantity of provisions, which were, of course, all lost. Here, then, the poor fellows found themselves, deprived of all the necessaries of life, and at a period of the year in which it was impossible to procure any wild fruit or roots. To return to the mountains was impossible, and their only chance of preservation was to proceed downwards, and to keep as near the banks of the river as circumstances would permit. The continual rising of the water had completely inundated the beach, in consequence of which they were compelled to force their way through an almost impervious forest, the ground of which was covered with a strong growth of prickly underwood. Their only nourishment was water, owing to which, and their weakness from fatigue and ill-health, their progress was necessarily slow. On the third day poor Maçon died, and his surviving comrades, though unconscious how soon they might be called to follow him, determined to keep off the fatal moment as long as possible. They therefore divided his remains in equal parts between them, on which they subsisted for some days. From the swollen state of their feet their daily progress did not exceed two or three miles. Holmes, the tailor, shortly followed Maçon, and they continued for some time longer to sustain life on his emaciated body. It would be a painful repetition to detail the individual death of each man. Suffice it to say that, in a little time, of the seven men, two only, named La Pierre and Dubois, remained alive. La Pierre was subsequently found on the borders of the upper lake of the Columbia by two Indians who were coasting it in a canoe. They took him on board, and brought him to the Kettle Falls, whence he was conducted to Spokane House."

" He stated that after the death of the fifth man of the party, Dubois and he continued for some days at the spot where he had ended his sufferings, and, on quitting it, they loaded themselves with as much of his flesh as they could carry; that with this they succeeded in reaching the upper lake, round the shores of which they wandered for some time in vain, in search of Indians; that their horrid food at length became exhausted, and they were again reduced to the prospect of starvation; that on the second night after their last meal, he (La Pierre) observed something suspicious in the conduct of Dubois, which induced him to be on his guard; and that shortly after they had lain down for the night, and while he feigned sleep, he observed Dubois cautiously opening his clasp knife, with which he sprang on him, and inflicted on his hand the blow that was evidently intended for his neck. A silent and desperate conflict followed, in which, after severe struggling, La Pierre succeeded in wresting the knife from his antagonist, and, having no other resource left, he was obliged in self-defence to cut Dubois's throat; and that a few days afterwards he was discovered by the Indians as before mentioned. Thus far nothing at first appeared to impugn the veracity of his statement; but some other natives subsequently found the remains of two of the party near those of Dubois, mangled in such a manner as to induce them to think that they had been murdered; and as La Pierre's story was by no means consistent in many of its details, the proprietors judged it advisable to transmit him to Canada for trial. Only one Indian attended; but as the testimony against him was merely circumstantial, and was unsupported by corroborating evidence, he was acquitted."

Meanwhile the greater part of this expedition continued their way through the mountains by the Athabasca Pass. Here, when surrounded by all the glory and grandeur of lofty mountains clad in eternal snow and icy glaciers, and amid the frequent crash and roar of descending avalanches, one of the *voyageurs* exclaimed, after a long period of silent wonder and admiration—" I 'll take my oath, my dear friends, that God Almighty never made such a place."

On the summit of the Athabasca Pass they were on the Atlantic side of the watershed, and here let us take leave of them while they pursue their toilsome journey across the great plains of Canada to the eastern side of the continent.

All of these early expeditions were undertaken in the interests of the fur trade, and carried out by the agents of the various fur companies, except for occasional bands of emigrants on their way to the Pacific Coast, the accounts of whose journeys are only referred to by later writers in a vague and uncertain manner.

The expedition in 1841 of Sir George Simpson, however, to which reference has been made in a previous chapter, is in many respects different from all the others. The rapidity of his movements, the great number of his horses, and the ease and even luxury of his camp life indicate the tourist and traveller, rather than the scientist, the hardy explorer, or the daring seeker after wealth in the wilderness. His narrative is the first published account of the travels of any white man in that part of the moun-

Discovery of Gold.   245

tains now traversed by the Canadian Pacific Road, though he mentions a party of emigrants which immediately preceded him in this part of his journey. The rapidity with which Sir George Simpson was wont to travel may be appreciated from the fact that he crossed the entire continent of North America in its widest part, over a route five thousand miles in length, in twelve weeks of actual travelling. The great central plains were crossed with carts, and the mountainous parts of the country with horses and pack-trains.

In 1858, gold was discovered on the upper waters of the Fraser River, and a great horde of prospectors and miners, together with the accompanying hangers-on, including all manner of desperate characters, came rushing toward the gold-fields, from various parts of Canada and the United States. This year may be considered as marking the birth of a new enterprise and the comparative decline of the fur trade ever after.

About this time, or, more precisely, in 1857, Her Majesty's Government set an expedition on foot, the object of which was to examine the route of travel between eastern and western Canada, and to find out if this route could be shortened, or in any other manner improved upon. Moreover, the expedition was to investigate the large belt of country, hitherto practically unknown, which lies east of the Rocky Mountains and between the United States boundary and the North Saskatchewan River. The third object of this expedition was to find a pass, or passes, available for horses across the Rocky

Mountains south of the Athabasca Pass, but still in British territory.

As this was an excellent opportunity for the advancement of science without involving great additional expense, four scientists, Lieut. Blackiston, Dr. Hector, Mr. Sullivan, and M. Bourgeau, were attached to the expedition. The party were under the control and management of Captain John Palliser.

The third object of this expedition is the only one that concerns the history of explorations in the Canadian Rockies. In their search for passes, Captain Palliser and Dr. Hector met with many interesting adventures, of which it is, of course, impossible to give more than the merest outline, as the detailed account of their journeys fills several large volumes. In August, 1858, Captain Palliser entered the mountains by following the Bow River, or south branch of the Saskatchewan. He then followed a river which comes in from the south, and which he named the Kananaskis, after an Indian, concerning whom there is a legend of his wonderful recovery from the blow of an axe, which merely stunned instead of killing him outright.

When they approached the summit of the pass, a lake about four miles long was discovered, round the borders of which they had the utmost difficulty in pursuing their way on account of the burnt timber, in which the horses floundered about desperately. One of the animals, wiser than his generation, plunged into the lake before he could be caught and proceeded to swim across. Unfortunately this animal was packed with their only luxuries, their tea, sugar, and blankets.

On the very summit of the pass is a small lake some half an acre in extent, which overflows toward the Pacific, and such was the disposition of the drainage at this point that while their tea-kettle was supplied from the lake, their elk meat was boiling in water from the sources of the Saskatchewan.

A few days later, Captain Palliser made a lone mountain ascent near one of the Columbia lakes, but was caught by night in a fearful thunder-storm so that he could not reach camp till next day. His descent through the forests was aided by the frequent and brilliant flashes of lightning.

A little later they met with a large band of Kootanie Indians, who, though very destitute and miserable in every other way, were very rich in horses. Captain Palliser exchanged his jaded nags for others in better condition, and despairing of pursuing his way farther, as the Indians were at war and would not act as guides, he started, on the first of September, to return across the mountains, and reached Edmonton in three weeks.

In the meantime Dr. Hector made a branch expedition which has some incidents of interest in connection with it. He was accompanied at first by the indefatigable botanist, M. Bourgeau, and by three Red River men, besides a Stoney Indian, who acted as guide and hunter for the party. Eight horses sufficed to carry their instruments and necessary baggage, as it was not considered necessary to take much provision in those parts of the mountains which he intended to visit.

Some reference has already been made to Dr. Hector's experiences in the vicinity of Banff, and we shall only give one or two of the more interesting details of his later travels. He left the Bow River at the Little Vermilion Creek, and followed this stream over the Vermilion Pass. The name of this pass is derived from the Vermilion Plain, a place where the ferruginous shales have washed down and formed a yellow ochre. This material the Indians subject to fire, and thus convert it into a red pigment, or vermilion.

Perhaps the most interesting detail of Dr. Hector's trip is that which occurred on the Beaverfoot River, at its junction with the Kicking Horse River. The party had reached the place by following down the Vermilion River till it joins the Kootanie, thence up the Kootanie to its source, and down the Beaverfoot. Here, at a place about three miles from where the little railroad station known as Leanchoil now stands, Dr. Hector met with an accident which gave the name to the Kicking Horse River and Pass. A few yards below the place, where the Beaverfoot River joins the Kicking Horse, there is a fine waterfall about forty feet high, and just above this, one of Hector's horses plunged into the stream to escape the fallen timber. They had great difficulty in getting the animal out of the water, as the banks were very steep. Meanwhile, Hector's own horse strayed off, and in attempting to catch it the horse kicked him in the chest, fortunately when so near that he did not receive the full force of the blow. Nevertheless, the kick knocked Hector down and rendered him senseless for some time. This was the more unfortunate, as they were out of food, and had seen no sign of game in the vicinity. His men ever after called the river the Kicking Horse, a name that has remained to this day despite its lack of euphony.

To the transcontinental traveller,

FALLS OF LEANCHOIL.

one of the most beautiful and inspiring points along the entire railroad is the descent of the Kicking Horse Pass from the station of Hector to Field. Here, in a distance of eight miles, the track descends 1000 feet, in many a curve and changing grade, surrounded by the towering cliffs of Mount Stephen and Cathedral Peak, while the rich forests of the valley far below are most beautiful in swelling slopes of dark green. Certainly, whoever has ridden down this long descent at breakneck speed, on a small hand-car, or railway velocipede, while the alternating rock cuts, high embankments, and trestles or bridges of dizzy height fly by in rapid succession, must feel at the same time a grand conception of the glories of nature and the triumphs of man. In striking contrast to this luxury of transportation was the old-time method of travelling through these mountains. The roaring stream which the railroad follows and tries in vain to descend in equally rapid slope is now one of the most attractive features of the scenery of the pass.

When Dr. Hector first came through this pass he had an adventure with one of his horses on this stream. They were climbing up the rocky banks of the torrent when the incident occurred. The horses had much difficulty in getting up, and in Hector's own words, " One, an old gray, that was always more clumsy than the others, lost his balance in passing along a ledge, which overhung a precipitous slope about 150 feet in height, and down he went, luckily catching sometimes on the trees ; at last he came to a temporary pause by falling right on his back, the pack acting as a fender. However, in his endeavors to get up, he started down hill again, and at last slid on a dead tree that stuck out at right angles to the slope, balancing himself with his legs dangling on either side of the trunk of the tree in a most comical manner. It was only by making a round of a mile that we succeeded in getting him back, all battered and bruised, to the rest of the horses."

That night they camped at one of the lakes on the summit of the pass, but were wellnigh famished. A single grouse boiled with some ends of candles, and odd bits of grease, served as a supper to the five hungry men.

The next day they proceeded down the east slope and came to a river that the Indian recognized as the Bow. About mid-day the Stoney Indian had the good fortune to shoot a moose, the only thing that saved the life of the old gray that had fallen down the rocky banks of the Kicking Horse River, for he was appointed to die, and serve as food if no game were killed that day.

Here we shall take leave of Dr. Hector and the Palliser expedition, and only briefly say that Hector followed the Bow to its source and thence down the Little Fork to the Saskatchewan and so out of the mountains. The next year Dr. Hector again followed up the Bow River and Pipestone River to the Saskatchewan, and thence over the Howse Pass to the Columbia, where he found it impossible to travel either west or northwest, and was forced to proceed southward to the boundary.

The main objects of the Palliser expedition were in a great measure accomplished, though the Selkirk Range of mountains was not penetrated by them, and no passes discovered through this formidable barrier. The vast amount of useful scientific material collected by the members of this expedition was published in London by the British Government, but it is now, unfortunately, so rare as to be practically inaccessible to the general reader.

The account of an expedition across the Rockies in 1862, by Viscount Milton and Dr. Cheadle, is perhaps the most interesting yet published. It abounds in thrilling details of unusual adventures, and no one who has read *The Northwest Passage by Land* will ever forget the discovery of the headless Indian when they were on the point of starvation in the valley of the North Thompson, or the various interesting details of their perseverance and final escape where others had perished most miserably. The object of this expedition was to discover the most direct route through British territory to the gold mines of the Caribou region, and to explore the unknown regions in the vicinity of the north branch of the Thompson River.

A period of very rapid growth in the Dominion of Canada now follows close upon the date of this expedition. In 1867, the colony of Canada, together with New Brunswick and Nova Scotia, united to form the new Dominion of Canada, and, in 1869, the Hudson Bay Company sold out its rights to the central and northwestern parts of British North America.

In the meantime the people of the United States had been vigorously carrying on surveys, and preparing to build railroads across her vast domains, where lofty mountain passes and barren wastes of desert land intervened between her rich and populous East and the thriving and energetic West, but in Canada no line as yet connected the provinces of the central plains with her eastern possessions, while British Columbia occupied a position of isolation beyond the great barriers of the Rocky Mountains.

On the 20th of July, 1871, British Columbia entered the Dominion of Canada, and on the same day the survey parties for a transcontinental railroad started their work. One of the conditions on which British Columbia entered the Dominion was, that a railroad to connect her with the east should be constructed within ten years.

More than three and one half millions of dollars were expended in these preliminary surveys, and eleven different lines were surveyed across the mountains before the one finally used was selected. Nor was this vast amount of work accomplished without toil and danger. Many lives were lost in the course of these surveys, by forest fires, drowning, and the various accidents in connection with their hazardous work. Ofttimes in the gloomy gorges and canyons, especially in the Coast Range, where the rivers flow in deep channels hemmed in and imprisoned by precipitous walls of rock, the surveyors were compelled to cross awful chasms by means of fallen trees,or, by drilling holes and inserting bolts in the cliffs, to cling to the rocks far above boiling cauldrons and seething rapids, where a fall meant certain death. The ceaseless exertion and frequent exposure on the part of the surveyors were often unrewarded by the discovery of favorable routes, or passes through the mountains. The Selkirk Range proved especially formidable, and only after two years of privation and suffering did the engineer Rogers discover, in 1883, the deep and narrow pass which now bears his name, and by which the railway seeks a route across the crest of this range, at the bottom of a valley more than a mile in depth.

The romance of an eagle leading to the discovery of a pass is connected with a much earlier date. Mr. Moberly was in search of a pass through the Gold Range west of the Selkirks, and one day he observed an eagle flying up a narrow valley into the heart of these unknown mountains. He followed the direction of the eagle, and, as though led on by some divine omen, he discovered the only route through this range, and, in perpetuation of this incident, the name Eagle Pass has been retained ever since.

But all these surveys were merely preliminary to the vast undertaking of constructing a railroad. At first, the efforts of the government were rewarded with only partial success, and at length, in 1880, the control and management of railroad construction was given over to an organization of private individuals. In the mountain region there were many apparently insuperable obstacles, to overcome which there were repeated calls for further financial aid. However, under the able and efficient control of Sir William Van Horne, the various physical difficulties were, one by one, overcome, while his indomitable courage and remarkable energy inspired confidence in those who were backing the undertaking financially. Moreover, he had a thorough knowledge of railroad construction, together with unusual perseverance and resolution, combined with physical powers which enabled him to withstand the nervous strain and worry of this gigantic enterprise.

In short, after a total expenditure of one hundred and forty million dollars, the Canadian Pacific Railroad, which is acknowledged to be one of the greatest engineering feats the world has ever seen, was completed, five years before the stipulated time.

With the opening of the railroad came the tourists and mountaineers, and the commencement of a new period in the history of the Canadian Rockies.

The short period of one hundred years which nearly covers the entire history of the Canadian Rockies may be divided into four divisions. The first is the period of the fur trade, which may be regarded as beginning with the explorations of Sir Alexander Mackenzie in 1793, and lasting till 1857.

From 1858 to 1871 might be called the gold period, for at this time gold-washing and the activity consequent upon this new industry were paramount.

The next interval of fifteen years might be called the period of railroad surveys and construction,—a time of remarkable activity and progress,—and which rationally closes in 1886, when the first trains began to move across the continent on the new line.

The last period is that of the tourists, and though as

yet it is the shortest of all, it is destined without doubt to be longer than any.

Every one of these periods may be said to have had a certain effect on the growth and advance of this region. The first period resulted in a greater knowledge of the country, and the opening up of lines of travel, together with the establishment of trading posts at certain points.

The second period brought about the construction of wagon roads in the Fraser Canyon leading to the Caribou mining region and to other parts of British Columbia. These roads were the only routes by which supplies and provisions could be carried to the mining camps. The method of gold mining practised in British Columbia has hitherto been mostly placer mining, or mere washing of the gravels found in gold-bearing stream beds.

With the commencement of the railroad surveys, a great deal of geographical information was obtained in regard to the several ranges of the Rocky Mountain system, and the culmination of this period was the final establishment of a new route across the continent, and the opening up of a vast region to the access of travellers.

Year by year there are increasing numbers of sportsmen and lovers of wild mountain life who make camping expeditions from various points on the railroad, back into the mountains, where they may wander in unexplored regions, and search for game or rare bits of scenery.

The future popularity of these mountains is in some degree indicated by the fact that those who have once tried even a brief period of camp life among them almost invariably return, year after year, to renew their experiences. The time will eventually come when the number of tourists will warrant the support of a class of guides, who will conduct mountaineers and sportsmen to points of interest in the wilder parts of the mountains, while well made roads will increase the comfort and rapidity of travel through the forests.

CHAPTER XV.

The Pleasures of the Natural Sciences—Interior of the Earth— Thickness of the Crust—Origin and Cause of Mountains—Their Age and Slow Growth—System in Mountain Arrangement—The Cordilleran System—The Canadian Rockies—Comparison with Other Mountain Regions— Climate—Cause of Chinook Winds—Effect of High Latitude on Sun and Moon—Principal Game Animals—Nature of the Forests—Mountain Lakes—Camp Experiences—Effect on the Character.

THOSE who have spent a few weeks or months in a mountain region, such as that of the Canadian Rockies, must soon come to feel an interest in those more striking features of the wilderness which have been constantly revealed. The special character of the mountains, which have given so much pleasure; the climate, on which, in a great measure, every action depends; the fauna, which adds so much of interest to the environment; and the flora, which increases the beauty of every scene—must all excite some degree of interest in those who have passed a short period of time surrounded by nature in her primeval state.

They spend their time to little advantage who do not thus become interested in the wonders of nature. A very slight knowledge of the habits and kinds of birds and animals, the principal characteristics of trees and plants,

Interior of the Earth. 259

the nature of minerals, the structure and formation of the earth's crust, and the laws which govern the circulation of currents in the atmosphere will, in every case, offer wide and boundless fields of research and pleasure. The camper, the huntsman, the explorer, and the mountaineer, armed with such information, will be prepared to spend the many hours of enforced idleness, which frequently occur by reason of fickle weather or a smoky atmosphere, in an interesting and profitable manner.

In the preceding chapters, the details of the flora and fauna, together with digressions on other topics, have been, from time to time, set forth in connection with various exploring excursions.

It is the purpose of this chapter, however, to discuss, in a general and very brief manner, such questions as have a special interest, and to present them in a somewhat more systematic manner than was possible, or natural, in connection with accounts of adventures.

To begin then with the foundation of things, the question first arises as to the origin and cause of mountains.

Astronomy teaches us that the earth is a mass of molten or semi-viscid matter, covered with a crust which has formed from the cooling of the exterior. As to the relative or absolute thickness of this crust, there is much diversity of opinion, but the great majority of estimates ranges between the limits of one hundred and one thousand miles.

The general features of the earth and the formation of mountains—subjects which lie in the province of geology

—likewise point to a comparatively thin crust covering a molten interior. Some geologists contend that the centre is likewise solid, and that there is a partially molten layer between the centre and crust. Now as the earth gradually cools by radiation, its volume diminishes, and the solid crust not having the strength to hold up its own weight, is forced to adapt itself to the contracting interior. The pressure thus brought to bear on the thin shell causes wrinkles or folds, so that the earth's surface is raised in some places and depressed in others. Moreover, the strata are folded, fractured, and thrown one over another as they are compressed, till at length lofty mountain ranges are formed, with all the phenomena of faults, flexures, and the wonderful contortions of the originally horizontal beds, that are to be observed in all mountain regions.

In some respects the mountains on the earth are comparable to the wrinkles on a drying apple, but in size, the highest peaks of the Himalayas and Andes have been compared more justly to the minute roughness on an egg shell.

Thus the mountain ranges of the world which appear so vast and lofty are exceedingly small and insignificant as compared with the great mass of the earth. The strength of the earth's crust seems incapable of supporting the weight of even these relatively small masses, for the highest peaks in the world never exceed an altitude of five and one half miles, a height which, if represented on a globe of ordinary size, would hardly be observable.

All the great mountain ranges of the world have been

raised to their present altitude since the Tertiary Age, but, nevertheless, we must conceive of mountain growth as a very slow and gradual process, a few feet or yards of elevation each century. That mountain chains have been upheaved at one or two violent convulsions of nature, is not in accordance with reason or geological facts. Faults are often found with a displacement of the strata through several thousand feet, a fact that has been used to prove a sudden catastrophe. But it should be held in mind that, after the strata were once fractured and made to slide one on another, the sliding would tend to be repeated at long intervals in this same place. Even then a yielding of but a few inches would be attended by a violent earthquake.

Beside the comparatively low altitude and very slow growth of mountain chains, there is a system in their arrangement which adds simplicity to the study of this subject. Dana calls attention to the fact that the great mountain chains of the earth are arranged along the borders of continents, and are proportional in height to the size of the oceans near them. The continents of North and South America reveal this law in a striking manner. The stupendous chain of the Andes in South America, and the more extensive Rocky Mountains in North America, stand opposite to the vast Pacific Ocean, and run nearly parallel to its shores, while the lesser systems on the eastern borders of each continent face the lesser area of the Atlantic Ocean. Moreover, almost all mountain chains show evidence of a pushing force from the direction of the sea, and a resisting force from the direction of the land.

The erosion of valleys commenced as soon as the strata were elevated above the sea-level, and thus the valleys of the world, being mostly those of erosion, are older than the mountains themselves.

Turning now to the Rocky Mountains or the Cordilleran System of North America, we observe that the chain extends from the region of the City of Mexico to the Arctic Ocean, and westward into the Alaskan Peninsula and the Aleutian Islands, a total distance of about five thousand miles. The Rocky Mountain system attains its greatest width in the latitude of Colorado, where it extends one thousand miles from east to west. Thence northward, the range becomes narrower toward the International boundary. From this point the system is only about four hundred miles in width, and the eastern range follows a line parallel to the Pacific Coast, nearly to the Arctic Circle.

Having thus very briefly glanced at the cause of mountain chains, the system in their arrangement, and the area covered by the Rocky Mountains of North America, let us turn our attention more particularly to the main features of the chain in its extension through Canada. In all, there are four ranges of mountains composing the Canadian Rockies. The most easterly is the highest and most important, and is, besides, the watershed between the Atlantic and Pacific drainage. Next to the west lie the Selkirk and Gold ranges, which must be grouped together. Near the Pacific Coast is a third range called the Coast Range, while Vancouver Island and the chain of islands extending north represent a fourth range of mountains. Between the two inner of these four ranges, there is a plateau region with an average altitude of 3500 feet.

Our attention centres with peculiar interest on the watershed or Summit Range, as in these mountains are found the grandest scenery and the most lofty peaks, and they are withal the most accessible to the traveller. On the eastern side, the Rocky Mountains rise abruptly from the plains and reach altitudes of 9000 to 11,000 feet. The plain is here, according to Dr. Dawson, about 4350 feet in altitude, while on the western side of the range the altitude of the Columbia valley is only 2450 feet, or nearly 2000 feet lower. The Summit Range is from forty to fifty miles wide in this portion of its course, and is made up of about five sub-ranges. The rivers and streams follow the valleys between these ranges, and find their way out of the mountains by occasional, transverse valleys, cutting through the ranges at right angles, so that every stream has a zig-zag course.

It would lead us too far to discuss the formations represented in the strata, and it is more important to learn the altitudes of the mountains above the valleys, and their other physical features, since these characteristics have a more direct bearing on the scenery and on the general nature of the mountains. The highest peaks of the Canadian Rockies rise from 5000 to 7000 feet above the valleys, and rarely surpass 11,000 or 12,000 feet altitude above sea-level. Thus they cannot compare in magnitude with the Himalayas, the Andes, or even the Swiss Alps. They, however, are more accessible than the Himalayas, are far more attractive than the Andes, and afford much greater variety of scenery, together with more beauty of vegetation, than the Alps. No picturesque hamlets adorn these valleys, no herds of cattle with tinkling bells pasture on these hillsides, and no well-made roads or maps guide the tourist to every point of interest; but, on the other hand, the climber may ascend mountains never tried before, the explorer may roam in wild valleys hitherto practically unseen by white men; and the camper may fish or hunt where no one besides the savage Indian has ever lowered a baited hook or joined in the stealthy chase.

Before leaving the discussion of geology, it would be well to call attention to the wonderful effects of ancient glacial action, everywhere in evidence among these mountains. The countless lakes were, almost without exception, formed in the Quaternary ice invasion. A few of the lakes occupy rock basins, and more are dammed by old terminal moraines, while the vast majority are held in by ridges of drift formed underneath the glaciers where they joined together at the confluence of valleys. Mention has already been made of the evidence of ice action on the summit of Tunnel Mountain, near Banff, showing that the ice was at least 1000 feet in thickness, but on the neighboring mountains there are further evidences that the ancient glaciers flooded this valley to a depth of 2700 or 2800 feet. Such evidences may be traced up the valley of the Bow to its source, where the upper surfaces of the glaciers were no less than 8500 or 9000 feet above sea-level, though these ice streams were about the same thickness as at Banff, because the valleys are much higher at this point. Throughout the eastern range, all the valleys were flooded, while only the mountain tops rose above the fields of ice, and the creeping glaciers moved slowly down the valleys and discharged in a great sheet of ice upon the plains to the east.

The climate of the Canadian Rockies is exceedingly cold in winter and temperate in summer, but the air is at all times so dry that changes of temperature are not felt as in lowland regions. The rainfall in summer is light, and rarely attended by heavy showers. The amount of snow and rainfall varies locally in a remarkable manner, by reason of the mountains themselves. Thus the maximum winter depth of the snow in the Bow valley may be two or three feet, when up in the higher regions, only five or six miles distant, the depth will approach fifteen or twenty feet. That mountains have a great influence on the climate and the amount of rainfall, is universally admitted. In fact, climate and mountains are mutually dependent one on the other. A range of mountains near the sea coast, if the circulation of the atmosphere carries the moist air over them, will cause a great precipitation of rain and snow, and, vice versa, the amount of precipitation decides the erosive power of streams, and consequently, the altitude and form of the mountains.

One of the most interesting features of the Canadian Rockies is the Chinook wind. These peculiar winds occur at all seasons of the year but are most noticeable in winter. At such times, after a period of intense frost, a wind springs up from the west, directly from the mountains, the temperature rises, and the snow disappears as if by magic. The air is so dry that the snow and moisture evaporate at once, leaving the ground perfectly free of moisture, where a few hours before was a deep covering of snow. Identical winds called Foehn winds occur in Switzerland, and in other mountain regions of the world. The explanation of these winds has been stated by Ferrel and others, but it is difficult of demonstration to those who do not understand the laws governing condensation and evaporation of moisture in our atmosphere. Most of these laws may be clearly illustrated by an experiment not very difficult to perform. A stout glass cylinder, closed at one end, is fitted with a closely fitting plunger. Now if a tuft of cotton, moistened with ether, be placed in the cylinder, and the plunger be suddenly and forcibly pushed in, the cotton will take fire. The compression of the air raises the temperature so that the cotton ignites. The experiment might have been reversed, and the plunger pulled suddenly outwards so as to rarefy the enclosed air. In this case the temperature of the air would have been much reduced, and, if there were sufficient moisture, it would condense on the sides of the cylinder or form a cloud of vapor. These experiments are exceedingly valuable, as they demonstrate the laws of temperature under changing pressure. Moreover, it shows how cold air discharges its moisture in the form of a mist, and thus illustrates the formation of the clouds in the upper cold regions of our atmosphere. Now the circulation of the air in the Canadian Rockies is, in general, from the Pacific Ocean across the mountains in an easterly direction. It is, of course, interfered with by the circular cyclonic storms which, from time to time, pass over the mountains. But when one or both causes of air motion compel the wind to blow from the west towards the east, the moist currents are forced to ascend and flow over the mountains. In this case the air becomes colder as it rises, mist and clouds are formed, and rain or snow falls, especially on the mountains themselves. As the air descends on the eastern side it becomes warmer in the increasing pressure, and the clouds evaporate and disappear. Now this air is much drier than when it left the other side of the mountains, because a great deal of rain and snow have been precipitated from it. Moreover, the latent heat given out as the clouds form, raises the temperature of the air above the normal temperature of those altitudes. This air gains heat as it descends, and is subjected to the increasing pressure of lower altitudes, and it finally appears as a warm and very dry wind on the east side of the mountains. Such a wind evaporates the snow, and causes it to disappear in a remarkably rapid manner.

The cause of Chinook winds is thus not difficult of explanation, if one understands the effects of atmospheric pressure and condensation. The latent heat given out by the condensing vapors and falling rain is of course equal to the heat furnished by the sun, when it was evaporating the surface waters of the ocean, and rendering the air full of invisible water vapor.

The aspect of the sky and clouds is one of the most beautiful features of the mountains. Except when obscured by the smoke of forest fires, the sky is at all times of that deep hue rarely seen near the sea-coast or in lowland regions. The dark blue extends without apparent paleness to the very horizon, while the zenith is of such a deep color, especially when seen from the summit of a lofty mountain, as to suggest the blackness of interstellar space. Against such a background, the brilliant cumulus clouds stand out in striking contrast, and every internal movement of the forming or dissolving vapors, as they rise, and descend, or curl about, is distinctly seen, because the clouds are so near.

The high latitude of this region has, of course, a considerable effect on the length of the days. Near the summer solstice the twilight is faintly visible all night, and the sun is below the horizon only a little more than six hours. The moon, however, is rarely visible in the summer months, because when near the full it occupies that part of the ecliptic opposite the sun, which, in this latitude, is much depressed. In consequence, the full moon runs her short arc so near the horizon that the high mountains shut out all view of her. In winter, these conditions are reversed, and the moon shines from the clear and frosty sky with unusual brilliancy, for many hours continuously, while the low-lying sun leaves many of the deeper mountain valleys without the benefit of his slanting rays for several months together.

It would be impossible to enumerate even the principal varieties of game animals, birds, and fish that inhabit this region. The mountain goat and sheep have been mentioned in previous chapters, and many of the interesting animals frequently met with have been described in more or less detail. The ordinary explorer or camper will see very little of the larger game, as he moves along with a noisy train of pack-horses and shouting men to drive them. He may occasionally see a bear, or catch sight of an elk or caribou, but the wary moose and the other members of the deer tribe will rarely or never be seen without an organized hunt. The camper will come to rely on the smaller game to give variety to his camp fare. Chief among these will be the grouse, of which there are six species in the Canadian Rockies. One variety is tame, or rather very stupid, and may be knocked down with stones, or snared with a strong elastic noose at the end of a pole. These birds are so numerous in the forests that one may always rely on getting a brace for dinner, after a little search, and I have even seen them walking about on the main street of Banff, where, of course, they are protected by law. Most of the mountain streams abound in trout, except where a high waterfall below has intercepted their coming up the stream. The larger lakes likewise afford fine fishing, and in many cases swarm with lake trout of a remarkable size. The camper will often obtain wild fowl, the black duck, mallards, and teal, in his excursions. Outside of these game birds and fish, there is little left for him to rely on, unless he chooses to dine on marmots and porcupines. These are often extolled by travellers as most excellent eating, but I have tried them both, and would prefer to leave my share to others, while there is anything else on hand.

The vegetation of the Canadian Rockies deserves a few remarks. The principal trees are all conifers. There are about six or seven species of these in the eastern range, and several more in the Selkirks. The paucity in the variety of deciduous trees in the Rocky Mountains, and the great number of conifers on the Pacific slope of North America, are in striking contrast to the wonderful number of deciduous species in the forests east of the Mississippi River. In the latter region, the number of species of forest trees is nowhere exceeded in the world, outside of tropical regions. Another remarkable fact in this connection was stated by Gray. He calls attention to the fact that there is a greater similarity, and affinity of species, between the Atlantic Coast trees and those of far distant Japan, than with those of the Pacific slope.

In the Canadian Rockies, trees cease to grow at altitudes above 7500 feet, under the most favorable circumstances, and the average tree line is in reality about 7000 feet. Bushes of the heath family and Alpine plants, however, reach much higher, while dwarfed flowering

herbs may be found in blossom as high as 8700 or 8800 feet. I once found a small mat of bright yellow sedums on the summit of a mountain, 9100 feet above sea-level, but this was an exceptional case. Above this altitude, various stone-gray, bright yellow, or red lichens, are the only sign of vegetable life. Nevertheless, in such cheerless regions of high altitudes, one sees a considerable variety of insect life—butterflies, wasps, mosquitoes, and spiders. The latter insects may sometimes be seen crawling about on the snow after winter has commenced, and naturalists have often described them as one of the most abundant insects on barren, volcanic islands of the Atlantic Ocean, where there is scarcely a trace of vegetation.

The pleasures of camping in the Canadian Rockies are almost infinite in their variety. They vary with the locality and the scenic interest of the surroundings, and suffer a constant change of mood and aspect with the changing weather. There is an exhilarating buoyancy in the mountain air that conspires to make all things appear as though seen through some cheerful medium, and where nature is so lavish with countless things of rare interest on every side, one comes at length to regard all other places unworthy of comparison. The formation of these mountains is such as to present an infinite variation of outline and altitude, such as one observes in almost no other mountain region of the world. The mountaineer may stand on the summit of a lofty peak and behold a sea of mountains extending fifty or one hundred miles in every direction, with no plains or distant ocean to suggest a limit to their extent. Such a vast area, nearly half a thousand miles in width, and thousands of miles in length, presents an extent of mountain ranges such as are found in no other part of the world.

The exquisite charm and beauty of the lakes, so numerous in every part of the mountains, is one of the chief delights of the camper. Some are small and solitary, perched in some amphitheatre far up among the mountains, surrounded by rocky walls, and hemmed in by great blocks of stone. Here, no trees withstand the Alpine climate, and the water surface is free of ice only during a short season. A few Alpine flowers and grasses wave in the summer breezes, while the loud whistling marmots, and the picas ever sounding their dismal notes, live among the rocks, and find shelter in their crevices.

Other lakes, at lower altitudes, are concealed among the dark forests, and, with deep waters, richly colored, appear like gems in their seclusion. Here the wild duck, the diver, and the loon resort in search of food, for the sedgy shores abound with water rice, and the waters with fish.

Most of the mountain lakes are small, and hide in secluded valleys, but many are large enough to become rough and angry in a storm, and have beaten out for themselves narrow beaches of gravel and shores lined with sand.

Even the sounds of the mountains and the forests give constant pleasure. There is every quality and volume of sound, from the loud rumble of thunder, or the

terrible crash of avalanches, re-echoed among the mountains, to the sharp, interrupted report of falling rocks, the roar of torrents, or the gentle murmur of some purling stream. The sighing of the wind in the forests, the susurrant pines and spruces, the drowsy hum of insects, the ripple of water on the shores of a lake, and the myriad sounds of nature—half heard, half felt—conspire to make up the sum of the camper's pleasure; though in a manner so vague and indescribable that they must needs be experienced to be understood.

Nor are all the experiences of camp life attended by pure enjoyment alone. Mountain adventures comprise a multitude of pleasures, mingled many times with disappointment and physical suffering. They comprise all the scale of sensations, from those marked by the pains of extreme exhaustion, physical weakness, hunger, and cold, to those of the greatest exhilaration and pleasure. Fortunately, the sensations of pleasure are by far the more abundant, while those of pain almost invariably follow some rash act or error in judgment.

The effect on the health and strength is, of course, one of the chief advantages of camp life. But there is another beneficial result brought about by this manner of life that is more important, though less often taken into consideration. This is the effect that camp life has on the character. In the first place, one learns the value of perseverance, for without this quality nothing can be accomplished in such a region as the Canadian Rockies. The explorer will realize this when he comes to a long stretch of burnt timber, where his horses flounder in a maze of prostrate trees; and the climber will feel the need of continued resolution when, after a long and arduous climb to an apparent summit, he reaches it only to find the slope extending indefinitely upwards.

The quality of patience under toil and aggravation while on the march—patience with tired horses and weary men—patience under the distress of wet underbrush, or uncomfortable quarters, or, indeed, when tormented by mosquitoes, is one of the prime requisites of life in the wilderness.

While these qualities are more or less common to every one, they are much developed in mountain camp life. But, perhaps, the ability to judge quickly and well is that characteristic which is most needed among the mountains, and the one which is attended by the most suffering if it is not brought into play. If the explorer or mountaineer decides on the time of day when he must turn back, and then, under the temptation of seeing a little more, or of reaching another summit, delays his return, let him not bewail his fate if he is caught by darkness in the forest and is compelled to pass a sleepless, hungry night. The laws of nature are inexorable, and while we obey them there is abundant opportunity of pleasure, but if we expose ourselves to the grinding of her vast machinery, one must suffer the consequence. The storm will not abate merely because we are exposed to it, nor will our strength be renewed merely because we are far from camp.

Camp Necessaries.

Let the camper surround himself with all the luxuries that are possible without trespassing on the bounds of reason. Let him have a good cook and a good packer; horses that are used to the trail; a fine camp outfit; comfortable blankets and good tents; a full supply of cooking utensils, knives, forks, and spoons; above all, let him take an abundant supply of provisions, comprising a large variety of dried fruits and the various cereals, and let each article be of the best quality.

Under such circumstances there is no risk of danger, no opportunity for discomfort, especially if every action is controlled by a moderate amount of judgment; but, on the other hand, the rich experiences among the mountains will prove a store of physical and mental resources, the memory of which will tempt him to revisit these regions year after year.

INDEX.

	PAGE
Abbott, Mount	130
Agnes, Lake	42
" " depth of	43
" " in winter	118
" " solitude of	42
Air circulation in Canadian Rockies	267
Alders in Selkirks	125
Alpine insects, varieties of	271
" plants	271
American Fur Company	238
Anemones	107
Assiniboine, another name for Stoneys	53
Assiniboine, Mount, altitude of	177
" " features of	178
" " first circuit of	168
" " " view of	153
" " outline	156
" " south side of	174
Astley, Mr.	61
Athabasca Pass	244
Atmosphere, eastward movement of	123
Avalanche from Mount Lefroy	33
Balsam fir	38
Banff, altitude	11
" climate	11, 15
" location	1
" population	2
" Springs Hotel	3
" surroundings	3
" topography of	4
Barometer, diurnal minima of	113
Bean, Mr.	72
Bear's Paw, chief of Stoneys	49
Beehive, the	41, 44
" altitude of	44

Index.

	PAGE
Blackiston, Lieut.	246
Blind valleys	172
Bourgeau, M.	246
Bow Lakes.	201
" " future popularity of.	211
" Lake, Lower	191
" " Upper.	195
" River.	2
British Columbia.	253
Brown, Mount, altitude of.	184
Bull-dog flies.	25
Butterflies, habits of.	72
Caledonia, New.	238
Calypso borealis.	143
Cambrian Age, reference to.	42
Canada, highest point reached in.	115
Canadian National Park.	1
" Pacific Road, cost of.	255
" Rockies, comparisons of.	264
Cannibalism, anecdote of.	242
Canoe River.	240
Caribou mining region.	256
Cascade Mountain, ascent of.	14
" " description of.	5
" " origin of name.	6
Castilleias.	107
Castle Crags.	65
Cave and basin at Banff.	159
Chalet at Lake Louise.	22
" old.	26
Character, effect of camp life on.	273
Cheops, Mount.	130
Chiniquy, Tom.	49
Chinook winds, cause of.	266
Chipmunks.	106
Cirque.	77
Climate of Canadian Rockies.	265
Cloud effects.	29, 52
Coast Range.	262
Condensation of clouds	267
Condition, physical.	89
Continental watershed.	18, 37
Contrast of surroundings.	95
Cold weather in September.	13
Colorado, altitude of mountains in.	37
Color, sunset and sunrise.	30

Index.

	PAGE
Columbia River	120
Columbine, yellow	19
Cook, Captain	230
" " explorations of	237
Cordilleran System	262
Coureurs des bois	221
Cox, Ross	240
Crees, Mountain	52
Crevasses, dangers of	203
Cross River	171
Daly Mountain	193
Dawson, Dr., on Stoney Indians	52
Desolation Valley	107
Devil's Club	125
" Head	7
" Lake	6
" " Indian legend of	8
Diamond hitch	142
Dominion of Canada	252
Eagle Pass	254
" Peak	126
" " later attempts on	129
Earth, interior of	259
Edith, Mount, Pass	219
Epilobium	107
Experiences in camp	273
Exploration, pleasure of	75, 96
Forbes, Mount, altitude of	184
Forest fires, ancient	188
" " causes of	188
" fire smoke	11
Forests, near Lake Louise	38
" of Pacific Coast	135
" regeneration of	190
" Selkirk	125
Forest trees, replacement of	190
Fraser, Simon	239
" River, first exploration of	238
Fur trade, origin of	220
Glacier, House	121
" debris	55
" thickness of ice in	78
Glissading, method of	70
Goat, Rocky Mountain	117, 163, 164
Gold, discovery of	245

Index.

	PAGE
Golden-rod, Alpine species	72
Gold Range	262
Great Mountain	77, 80
" Slave Lake, origin of name	231
Green, Dr.	124
Grouse	269
Hazel Peak, altitude	108
Hector, Dr.	10, 246
Hector, Mount	216
Heely's Creek	139
Hermit Range	127
Hooker, Mount, altitude of	184
Huber and Sulzer	124
Hudson Bay Company	224, 226
Ice Age	5, 264
" pillars	56
Indian, ability to follow trails	49
" frankness	63
" gratitude	51
" habits of	50
" idea concerning weeping	234
" Kootanie	247
" loquacity	8
" pathos	51
" sarcasm	100
" summer	29
" trails	152, 212
Kananaskis Pass, legend of	246
Kicking Horse Pass, discovery of	250
" " River, origin of name	249
Kootanie River, direction of flow	120
Laggan, distance from Banff	62
" distance to Lake Louise	24
" " " Mount Temple	79
Lake Louise, altitude of	22
" " depth of	17
" " early morning at	26
" " forests about	23
" " highest recorded temperature at	22
" " in October	31
" " past history of forests at	24
" " prevalent wind at	24
" " size and shape of	16

Index.

	PAGE
Lake Louise, summer temperature of water	26
" " topography of region near	36
" " visitors at	22
Lakes in Canadian Rockies	272
Lake trout, size of	6, 202
Laurel, sheep	19
Lefroy, Mount, avalanche from	90
" " description of	18
" " precipices of	34, 90
Linnea borealis	175
Little Fork Pass, altitude of	208
Lyall's larch	39
Mackay, Alexander	232
Mackenzie, Sir Alexander	226
" River, discovery of	231
Mackenzie's plan for an overland route	238
Maple trees	175
Marion Lake	130
Mariposa Grove	135
Marmots	43, 106
Milton and Cheadle	252
Minerals on mountain sides	173
Minnewanka Lake	8
Mirror Lake	42, 45
Moon, effect of latitude on	268
Morley	51
Mosquitoes, annual disappearance of	25, 199
Mountains, age of	261
" altitude of	263
" comparative size of	260
" origin and cause of	259
" system in arrangement	261
Mountaineers, tribulations of	113
Mounted Police, Northwest	2
Murchison, Mount	207
Muskegs	46
Névé regions	56
Northwest Company	224
" " downfall of	225
" Mounted Police	2
Outfit for camp	275
Pacific Coast reached by Mackenzie	234
Pack-horses, difficulties with	102
" " nature of	214
" " remarkable experience with	105

Index.

	PAGE
Palliser expedition	245
Paradise Valley	91, 105
" " discovery of	91
" " in winter	117
" " location	84
Patience, need of, in camp life	274
Peace River, origin of name	231
Peechee	8
Perseverance, need of, in camp life	274
Peyto, William	140
Phlox, alpine	72
Pica, tailless hare	105
Pinnacle Mountain	92, 97
Plateau region	263
Pleasure of camp life	271
Ptarmigan	76
Quesnel, Jules	238
Rat, wood	106
Rhododendron	20
Roger's Pass	254
Rundle Mountain	5
Rundle's early visit to Banff region	9
Rundle the missionary	9
Rupert, Prince	224
Saddle, the	77
Saskatchewan	206
Scenery at high altitudes	45, 89, 108
Schrunds, cause of	87
Selkirk Range, humidity of	123
Selkirks, early popularity of	124
" forest trees	132
" geographical position	119
Sheep, mountain	10
Simpson Pass	144
" River	145
" Sir George	244
Simpson's, Sir George, expedition	7
Sky, color of, in mountains	268
Smoke of forest fires	81
Snow line, determination of	56
" " in Colorado and the Andes	57
" patches, effect on vegetation	67
" storm in June	12
Solitude of high altitudes	215
Sounds, forest and mountain	273

Index. 283

	PAGE
Spray River	4
Spruce trees	38
Stones, loose, danger of, in Canadian Rockies	59, 69
Stoney Indians, characteristics of	51
" " dress of	52
" " nature of	9
" " Palliser's account of	53
" " religion of	52
Storms, approach of	27, 81
" mountain	156
St. Piran, flowers and butterflies	72, 73
" summit of	71
Stuart, John	239
Sullivan, Mr	246
Sulzer and Huber	124
Surveys for railroad	253
Temple, Mount, altitude	78
" " first ascent of	115
" " avalanches from	78
" " first attempt to ascend	109
" " maximum temperature on summit	116
" " north side of	110
" " strata of	79
" " summit of	116
Thirst, method of quenching	74
Thunderstorms in mountains	28
Tourists at Banff	2
Tree line	270
Trees, age of	135
" " at Lake Louise	23
Tunnel Mountain	4
Twilight, length of	11
Twin, William	48
Valleys, age of	262
" blind	212
Van Horne, Sir William	254
Vegetation of Canadian Rockies	270
Vermilion Pass	181
" Plain	248
" River	179
Vitality of mountain trees and herbs	12
Voyageurs	221
Waputehk Range	183
Wasps and bull-dog flies	25
Wildman, Enoch	48, 119
Wilson, Tom	138

www.ingramcontent.com/pod-product-compliance
Lightning Source LLC
Chambersburg PA
CBHW030001240426
43672CB00007B/777